KB074049

What Koreans Eat

What Koreans Eat

by Kyung Shin

illustrated by Kyung Shin

G-world

What Koreans Eat

by Kyung Shin

illustrated by Kyung Shin

Copyright @ 2024 by Kyung Shin

First printing, published 22, April, 2024

Author	Kyung Shin
Publisher	Kibong Lee
Editor	G-world editing team
Publisher	G-world Publishing Co.
Address	G-World Building, Yangwha-ro 12 Road 26, Mapo-gu, Seoul, Korea
Tel.	02) 374-8616~7
Fax.	02) 374-8614
Email	gworldbook@naver.com
Homepage	www.g-world.co.kr

ISBN   979-11-388-3019-5 (03590)

Foreword

Dear Kyung,

Congratulations on the publication of your artistic cookbook! Your creative spirit shines through every page, making this book a true reflection of your passion for both art and Korean cuisine.

Before my tenure as a professor in the Department of Food and Nutrition at Daejeon University, I spent almost a decade living abroad. In those challenging times, self-made Korean food became my solace, offering a balm for mental fatigue. It never crossed my mind, during my years of lecturing, to compile a cookbook, but you, dear Kyung, embraced this endeavor with grace and vision.

Our journey dates to 1972 when we first met as freshmen at Seoul National University's College of Human Ecology, Department of Food and Nutrition. I recall your artistic prowess even then, with your skillful figure drawings leaving an impression. Throughout college, we shared the trials of exams, the rigors of experimental reports, and engaged in volunteering activities, creating a tapestry of shared experiences.

In 1986, life led you to the United States, where you evolved into a talented painter. For the past 13 years, you've actively participated in exhibitions, weaving your artistic journey with dedication.

Your dedication to food, rooted in your academic background, manifested in 2015 when you started a blog sharing Korean-food recipes alongside your paintings. Responding to requests from your American friends, you ventured into a second blog, featuring recipes in English alongside your recent artwork. Now, with your unique cookbook, you beautifully blend your culinary expertise, nutritional insights, and artistic talents.

The cookbook's 200 recipes, spanning various categories, are accompanied by illustrations for clarity. Notably, you've tailored the recipes for non-experts, using only Korean ingredients available in the US. The inclusion of Koreanized versions of Japanese, Chinese, and Western dishes add a distinctive touch.

This cookbook, beyond being a collection of recipes, is a repository of memories. As I peruse its pages, I'm reminded of your mother, mother-in-law, family, and relatives. The warmth, love, and care experienced during meals at your home echo through these dishes. Cooking Korean food may pose challenges, but the rich flavors and essential nutrients contribute to both physical and mental well-being.

As I write this on January 10th, 2024, from Seoul, Republic of Korea, I am filled with pride as a long-time friend. I wholeheartedly believe that your cookbook will not only preserve Korean culinary traditions but also serve as a bridge, passing on this cultural richness to second and third-generation Korean Americans.

Warm regards,

Nan-Sook Koo

To the Readers

Hi!

Nearly 38 years have passed since I first set foot on American soil. My journey began as a dedicated dietitian in Korea, where I spent a substantial 5 years at ADD. Fast forward to my early 50s, and I immersed myself in American culinary culture, embracing the role of a dietitian. It's worth noting that with completion of DPD, I am not a Registered Dietitian (RD) due to unforeseen circumstances during the internship stage.

As life, ever unpredictable, led me down the path of becoming an artist, it brought a new chapter to my life. Yet, amidst the strokes of creativity, my passion for cooking remains unwavering. When the pressures of artistry become too much, the kitchen is my sanctuary—where I cook away stress and rediscover joy.

My dedication to culinary education took on a new dimension as I engaged in various volunteer roles as a cooking class instructor. Picture this; guiding adopted children through the delightful world of cooking, crafting specialized dishes for moms of children with autism, and sharing the art of Kimchi-making with my high school alma mater and the vibrant community of women at my church.

As an adopted member of the so-called salad bowl that is America, I've assimilated a myriad of ethnic flavors into my daily meals. With the increasing interest in K-food, it feels like the perfect time to share my recipes.

This book is more than a collection of recipes; it's a narrative of my journey and a celebration of diverse culinary experiences. Join me in savoring the richness of our food culture and extending a warm welcome to others.

With culinary wishes,

Kyung Shin

Author and Culinary Explorer

Contents

Foreword   • 6
To the Readers   • 8

Introducing Recipes / 200 recipes
Main Dishes / 54 recipes   • 13
Side Dishes / 45 recipes   • 15
Rice, Porridge & Noodles / 24 recipes   • 16
Soups & Stews / 26 recipes   • 17
Kimchi & Pickles / 22 recipes   • 18
Desserts / 29 recipes   • 19

Chapter I   Main Dishes   • 21
Chapter II   Side Dishes   • 93
Chapter III   Rice, Porridge & Noodles   • 145
Chapter IV   Soups & Stews   • 185
Chapter V   Kimchi & Pickles   • 221
Chapter VI   Desserts   • 257

Glossary   • 301
Index   • 304
Searching in Korean by Korean Alphabetical Orders   • 311
Acknowledgement   • 313
Reference   • 314

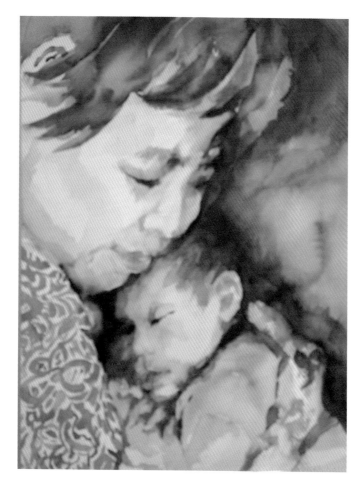

Artist: Kyung Shin
Title: Love
Size: 14"×20"
Medium: watercolor
North East Watercolor Society 46th Annual International Exhibition-2022, Kent, NY-exhibitor

Introducing Recipes

<Main Dishes>

Baked Croaker (Jogi Gui / 조기구이) • 23

Batter-fried Chives (Buchujeon / 부추전) • 24

Batter-fried Napa Cabbage (Baechujeon / 배추전) • 25

Batter-fried Stuffed Green Pepper (Putgochujeon / 풋고추전) • 26

Batter-fried Stuffed Mushrooms (Beoseot Jeon / 버섯전) • 27

BBQ Beef Short Ribs (LA Galbi / 엘에이갈비) • 28

Beef Braised in Soy Sauce (Jangjorim / 장조림) • 30

Braised Beef Short Ribs (Galbijjim / 갈비찜) • 31

Braised Chicken (Dakjjim / 닭찜) • 33

Braised Croaker (Jogijjim / 조기찜) • 34

Braised Fishcake (Eomuk Bokkeum / 어묵볶음) • 35

Braised Half-dried Pollock (Kodari Jjim / 코다리찜) • 36

Braised Mackerel (Godeungeo Jorim / 고등어조림) • 37

Braised Stalked Sea Squirts (Mideodeok Jjim / 미더덕찜) • 38

Broiled Pork Ribs (Dwaeji Galbi / 돼지갈비) • 40

Bulgogi (불고기) • 42

Cabbage Rolls (Yangbaechu Roll / 양배추롤) • 43

Fried Cuttlefish (Ojingeotwigim /오징어튀김) • 44

Hamburger Patties (햄버거패티스) • 46

Hand Rubbed Beef (Jumulleok / 주물럭) • 48

King Dumpling (Wang Mandu / 왕만두) • 49

Korean Patties (Wanjajeon / 완자전) • 51

Korean Style Egg Roll (Gyeran Mari / 계란말이) • 52

Lasagna • 53

Mapo Tofu (Mapa Dubu / 마파두부) • 55

Meat Sauce for Spaghetti in Crock Pot • 57

Mung Bean Pancake (Bindaetteok / 빈대떡) • 58

Pan-fried Fish Fillet (Saengseonjeon / 생선전) • 60

Pan-fried Slice of Beef (Yukjeon / 육전) • 61

Passover Broiled Beef Flank • 62

Platter of Nine Delicacies (Gujeolpan / 구절판) • 63

Pork Bulgogi (Dwaeji Bulgogi / 돼지불고기) • 65

Seafood Dynamite • 66

Seafood Scallion Pancake (Pajeon / 파전) • 67

Seasoned Dried Pollock (Bugeopo Muchim / 북어포무침) • 68

Seasoned Shredded Squid (Ojingeo Chae Bokkeum / 오징어채볶음) • 69

Spicy Braised Chicken (Dakdoritang / 닭도리탕) • 70

Spicy Braised Tofu (Dubu Jorim / 두부조림) • 71

Steamed Egg (Gyeran Jjim / 계란찜) • 72

Stir-fried Anchovy (Myeolchi Bokkeum / 멸치볶음) • 73

Stir-fried Dried Pollock (Bugeopo Bokkeum / 북어포볶음) • 74

Stir-fried Dry Squid Slice (Mareun Ojingeo Bokkeum / 마른오징어볶음) • 75

Stir-fried Glass Noodles & Vegetables (Japchae / 잡채) • 76

Stir-fried Glass Noodles with Bell Peppers (Pimang Japchae / 피망잡채) • 78

Stir-fried Glass Noodles with Mushrooms (Beoseot Japchae / 버섯잡채) • 80

Stir-fried Golden-haired Squid (Ojingeo Silchae Bokkeum / 오징어실채볶음) • 82

Stir-fried Jiri Anchovies (Jiri Myeol Bokkeum / 지리멸볶음) • 83

Stir-fried Octopus (Nakji Bokkeum / 낙지볶음) • 84

Stir-fried Oyster Mushrooms (Neutaribeoseot Bokkeum / 느타리버섯볶음) • 86

Stir-fried Pork Belly (Samgyeopsal Bokkeum / 삼겹살볶음) • 87

Stir-fried Rice Cake (Tteokbokki / 떡볶이) • 88

Stir-fried, Yellow-dried Pollock (Hwangtaegui / 황태구이) • 89

Sweet and Sour Pineapple Pork (Tangsuyuk / 탕수육) • 90

Sweet and Sour Wings • 92

<Side Dishes>

Acorn Jelly Making (Dotori Muk Ssugi / 도토리묵쑤기) • 95

Black Beans with Soy Sauce (Kongjorim / 콩조림) • 96

Braised Bellflower Roots (Doraji Namul / 도라지나물) • 97

Braised Bracken (Gosari Namul / 고사리나물) • 98

Braised Burdock Root (Ueong Jorim / 우엉조림) • 99

Braised Lotus Roots (Yeongeun Jorim / 연근조림) • 100

Chayote Salad (Chayote Muchim / 차요테무침) • 101

Chicken Gravy • 102

Chives Salad (Buchu Muchim / 부추무침) • 103

Cold Cucumber Seaweed Soup (Oi Miyeok Naengguk / 오이미역냉국) • 104

Egg Roll Slices (Hwangbaekjidan / 황백지단) • 105

Fermented Dry Squid Slice with Rice (Bap Sikhae / 밥식해) • 106

Fermented Squid with Radish (Ojingeo Sikhae / 오징어식해) • 107

Fried Gim with Rice Paper (Gim Bugak / 김부각) • 108

Fried Kelp (Dasima Twigak / 다시마튀각) • 109

Gim Mix (Gim Muchim / 김무침) • 110

Ginger Dressing • 111

Greek Yogurt with Cucumber & Apple • 112

Green Garlic Stem Dish (Putmaneuldaemuchim / 풋마늘대무침) • 113

Jellyfish Salad (Haepari Naengchae / 해파리냉채) • 115

Korean Style Dipping Sauce (Ssamjang / 쌈장) • 117

Kyung's Apple Salad • 118

Pan-fried Eggplant (Gaji Gui / 가지구이) • 119

Parboiled Spinach Dish (Sigeumchi Namul / 시금치나물) • 120

Radish Salad with Sugar & Vinegar (Mu Saengchae I / 무생채 I) • 121

Radish Salad without Sugar & Vinegar (Mu Saengchae II / 무생채 II) • 123

Raw Crab Marinated in Soy Sauce (Ganjang Gejang / 간장게장) • 124

Scallion Salad (Pamuchim / 파무침) • 126

Seasoned Crab (Yangnyeom Gejang / 양념게장) • 127

Seasoned Cucumber and Bellflower Roots (Oi Doraji Muchim / 오이도라지무침) • 129

Seasoned Dried Radish (Mu Mallengi Muchim / 무말랭이무침) • 130

Seasoned Mung Bean Sprouts (Sukju Namul Muchim / 숙주나물무침) • 131

Seasoned Soybean Paste (Gang Doenjang / 강된장) • 132

Seasoned Soybean Sprouts (Kongnamul Muchim / 콩나물무침) • 133

Steamed Eggplant (Gaji Namul / 가지나물) • 134

Stir-fried Chayote (Chayote Bokkeum / 차요테볶음) • 135

Stir-fried Eggplant (Gaji Bokkeum / 가지볶음) • 136

Stir-fried Eggplant & Anchovy (Gaji Myeolchi Bokkeum / 가지멸치볶음) • 137

Stir-fried Radish (Mu Namul / 무나물) • 138

Stir-fried Red Pepper Paste (Bokkeum Gochujang / 볶음고추장) • 139

Stir-fried Soybean Sprouts (Kongnamul Bokkeum / 콩나물볶음) • 140

Stir-fried Zucchini (Hobak Bokkeum / 호박볶음) • 141

Vinegary Cucumber Side Dish (Oi Chomuchim / 오이초무침) • 142

Vinegary Lotus Roots (Yeongeun Chojeolim / 연근초절임) • 143

White Radish Pickle (Tongdak Jipmu / 통닭집무) • 144

<Rice, Porridge & Noodles>

Abalone Porridge (Jeonbokjuk / 전복죽) • 147

Beef rice Bowl (Gogi Deopbap / 고기덮밥) • 148

Bibimbap (비빔밥) • 149

Black Bean Sauce Noodles (Jjajangmyeon / 짜장면) • 150

Chinese Udong (Junghwa Udong / 중화우동) • 152

Curry Rice (카레라이스) • 154

Eggplant Rice Bowl (Gaji Deopbap / 가지덮밥) • 155

Gimbap (김밥) • 156

Knife-cut Noodle Soup with Seafood (Kalguksu / 칼국수) • 158

Korean Ginseng Chicken Soup (Samgyetang / 삼계탕) • 160

Mixed Cold Buckwheat Noodles (Bibim Naengmyeon / 비빔냉면) • 161

Multi-grain Rice (Japgokbap / 잡곡밥) • 162

New Year's Day Rice Cake Soup (Tteokguk / 떡국) • 164

Omurice (오므라이스) • 166

Pesto Sandwiches • 168

Pine Nut Porridge (Jatjuk / 잣죽) • 169

Rice with Kimchi & Soybean Sprouts (Kimchi Kongnamulbap / 김치콩나물밥) • 170

Rice with Soybean Sprouts (Kongnamulbap / 콩나물밥) • 172

Spicy Korean Acorn Noodles (Bibim Dotori Guksu / 비빔도토리국수) • 174

Spicy Seafood Noodle Soup (Jjamppong / 짬뽕) • 176

Sushi Rice • 178

Tuna Sashimi Rice Bowl (Hoedeopbap / 회덮밥) • 179

Udong (Udong / 우동) • 181

Watery Cold Buckwheat Noodles (Mul Naengmyeon / 물냉면) • 182

<Soups & Stews>

Bean Curd Stew (Kongbijijjigae / 콩비지찌개) • 187

Beef Bone Soup (Sagol Guk / 사골국) • 189

Brisket Vegetables Rice Soup (Gari Gukbap / 가리국밥) • 191

Chicken Corn Soup, Chinese Style • 193

Clear Stew with Codfish Head (Daegu Meoritang / 대구머리탕) • 194

Cold Gim Soup (Gim Guk / 김국) • 195

Corn Cream Soup • 196

Crabmeat Soup (Gesal Soup / 게살수프) • 197

Fishcake Soup (Eomukttang / Odeng / 어묵탕, 오뎅) • 198

Hand Torn Dough Soup (Sujebi & Gamja Ongsimi / 수제비와 감자옹심이) • 199

Kimchi Stew (Kimchi Jjigae / 김치찌개) • 200

Mixed Bone Soup (Seolleongtang / 설렁탕) • 201

New England Clam Chowder • 203

New York Style Clam Chowder • 204

Oxtail Soup (Kkori Gomtang / 꼬리곰탕) • 205

Radish Beef Soup (Mu Guk / 무국) • 206

Rich Soybean Paste Stew (Cheonggukjang Jjigae / 청국장찌개) • 207

Salted Pollock Roe Stew (Myeongran Jjigae / 명란찌개) • 208

Sesame Dried Pollock Soup (Bugeotguk / 북엇국) • 209

Seaweed Soup (Miyeok Guk / 미역국) • 210

Soft Tofu Stew (Sundubu Jjigae / 순두부찌개) • 211

Soybean Paste Soup with Spinach & Soybean Sprout (Doenjangguk / 된장국) • 213

Soybean Paste Stew (Doenjang Jjigae / 된장찌개) • 214

Spicy Beef & Vegetable Soup (Yukgaejang / 육개장) • 215

Vegetable Soup (Yachae Soup / 야채수프) • 217

Wild Pollock Stew (Maeuntang / 매운탕) • 218

<Kimchi & Pickles>

Boiled Pickled Cucumber (Oi Sukjangajji / 오이숙장아찌) • 223

Cabbage Kimchi (Yangbaechu Kimchi / 양배추김치) • 224

Chayote Pickle (Chayote Jangajji / 차요테장아찌) • 225

Chives Kimchi (Buchu Kimchi / 부추김치) • 226

Cucumber Pickle (Oiji / 오이지) • 227

Fresh Baby Napa Kimchi (Eolgari Geotjeori / 얼가리겉절이) • 229

Fresh Baechu Kimchi (Baechu Geotjeori / 배추겉절이) • 231

Fresh Chives Kimchi (Buchu Geotjeori / 부추겉절이) • 233

Fresh Tomato Kimchi (Tomato Geotjeori / 토마토겉절이) • 234

Pickled Green Pepper (Putgochu Jangajji / 풋고추장아찌) • 235

Pre-cut Kimchi (Mak Kimchi / 막김치) • 236

Radish Kimchi (Kkakdugi / 깍두기) • 238

Spicy Stuffed Cucumber Kimchi (Oi Sobagi / 오이소박이) • 240

Sauerkraut • 242

Watery Cabbage Kimchi (Yangbaechu Mul Kimchi / 양배추물김치) • 243

Watery Cucumber Kimchi (Oi Mul Kimchi / 오이물김치) • 244
Watery Radish Kimchi (Mu Mul Kimchi / 무물김치) • 246
Watery Sedum Kimchi (Dolnamul Mul Kimchi / 돌나물물김치) • 247
White Kimchi (Baek Kimchi / 백김치) • 248
Whole Kimchi (Baechu Kimchi / 배추김치) • 250
Whole Radish Kimchi (Chonggak Kimchi / 총각김치) • 252
Young Radish Kimchi (Yeolmu Kimchi / 열무김치) • 254

\<Desserts\>

Baked Sweet Potatoes in Air Fryer • 259
Bean Juice (Kongguk / 콩국) • 260
Carrot Patties • 261
Cinnamon Ginger Punch (Sujeongwa / 수정과) • 262
Cinnamon Spiced Sweet Potatoes • 264
Cranberry Sauce • 265
Cream Puffs • 266
Deep-fried & Sugar Glazed Banana • 267
Deep-fried & Sugar Glazed Sweet Corn Ball • 269
Fruit Compote / Fruit Nut & Rice Salad • 271
Glutinous Rice Cake (Injeolmi / 인절미) • 272
Korean Sweet Dessert Pancakes (Hotteok / 호떡) • 274
Korean Sweet Potato Snack (Matang / 마탕) • 276
Korean Sweet Rice Cake with Dried Fruit and Nuts (Yak Sik / 약식) • 277
Mochi • 278
Namagashi (Saenggwaja / 생과자) • 279
Organic Green Smoothie (Yachae Smoothie / 야채스무디) • 281
Pine-needle Rice Cake (Songpyeon / 송편) • 282
Pizzelle Cookies • 284

Potato Pancake (Gamjajeon / 감자전) • 286

Red Bean Sediment (Pat Anggeum / 팥앙금) • 287

Rice Cake Ball (Gyeongdan / 경단) • 288

Rice Punch (Sikhye / 식혜) • 289

Scones • 291

Sorghum Balls (Susu Pat Danji or Susu Gyeongdan / 수수팥단지, 수수경단) • 293

Sticky Rice Cake, Covered by Red Beans (Siru Tteok / 시루떡) • 295

Tiramisu / Ladyfinger Coffee Cake • 297

Traditional Korean Sweet Pastry (Yakgwa / 약과) • 299

Twisted Cookies (Maejakgwa / 매작과) • 300

Chapter 1

Main Dishes

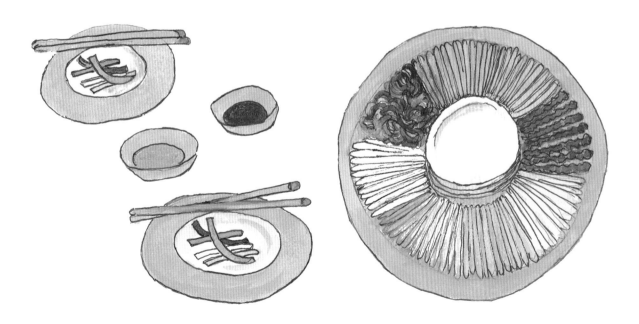

Baked Croaker (Jogi Gui / 조기구이)

Total Time: 1 hour

Ingredients (for 4 persons):
- 4 yellow croakers (10 ounces each)
- ¼ cup of coarse sea salt
- 2 tablespoons of roasted sesame oil (for coating)

Method:

1. If possible, purchase already cleaned croakers to save time and effort. However, if you have fresh croakers, descale them, paying extra attention to descaling around the head area; other parts are fine. Use scissors to trim fins and the tail and remove the innards. Rinse the croakers with cold running water and drain them. Set them aside.

2. Sprinkle the croakers with coarse sea salt for about 30 minutes. Coarse sea salt will firm up the skin of the croakers and retain their savory juices without making them too salty.

3. Preheat the oven to 390 degrees Fahrenheit. Place the brined croakers on a baking dish and bake them for approximately 30 minutes.

Enjoy your delicious roasted yellow croakers!

Batter-fried Chives (Buchujeon / 부추전)

Total Time: 20 minutes

Ingredients (for 4 persons):
- ¼ bundle of chives (approximately 5 ounces), whole
- ½ cup wheat flour
- 1 teaspoon of Kosher salt
- 1 small red pepper chopped and sliced (1 ounce)
- 1 egg, beaten (optional)
- ⅓ cup of water
- 2 tablespoons of grapeseed oil

Vinegary Soy Sauce:
- 2 tablespoons of soy sauce
- 1 tablespoon of rice vinegar
- ½ teaspoon of sesame seeds
- ½ teaspoon of organic sugar

Method:

1. Cut the 1-inch white part of the chives, trim the whole leaves, wash, and drain. Set them aside.
2. Prepare the batter by mixing flour, salt, green pepper, egg and water. Gently fold the mixture; avoid overmixing, as it can make the batter sticky and result in less crispiness.
3. Heat a frying pan over strong heat, add grapeseed oil, and pour the batter mixture onto the pan. Place the chives on top of the batter. Reduce the heat to medium and keep cooking until the batter turns its color to golden-yellow, about 5 minutes, flipping halfway.
4. Serve with vinegary soy sauce for dipping.

Tip:

It's recommended not to cut the chives but use them whole. This recipe originates from Gyeongsang Province and is known for its unique and delicious batter-fried chives, as taught by my mother-in-law. You can omit the egg for a softer texture, as per the traditional method.

Batter-fried Napa Cabbage (Baechujeon / 배추전)

Total Time: 20 minutes

Ingredients (for 2 persons):
- 4 medium-sized whole Napa cabbage leaves
- ½ cup of wheat flour
- 1 teaspoon of Kosher salt
- ½ cup of water
- 2 tablespoons of grapeseed oil

Dipping Sauce:
- 2 tablespoons of soy sauce
- 1 tablespoon of rice vinegar
- 1 teaspoon of red pepper powder
- ½ teaspoon of minced garlic
- 1 teaspoon of organic sugar
- ½ teaspoon of roasted sesame seeds
- 1 teaspoon of sesame oil

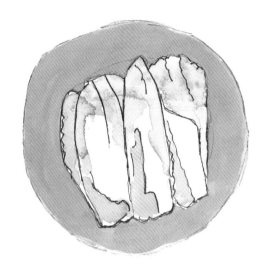

Method:

1. Wash the 4 sheets of Napa cabbage leaves and allow them to drain. Set it aside.
2. Gently knock and press the thick stalk part of the cabbage leaves with the back of a knife once or twice.
3. Prepare the batter by mixing flour, salt, and water. Dip each cabbage leaf one by one into the batter.
4. Heat a frying pan over strong heat, add grapeseed oil, and stir-fry the cabbage leaves dipped in the batter, making sure to cook both sides. Reduce the heat to medium and cook for about 5 minutes till the batter turns its color to golden-yellow, flipping halfway.

Tip:

I prefer not to use pre-packaged batter mixes, as they often contain added sugars, salt, baking soda, and sometimes preservatives.

Batter-fried Stuffed Green Pepper (Putgochujeon / 풋고추전)

Total Time: 1 hour

Ingredients (for 4 persons):
- 4~6 green peppers (4~6 ounces)
- 4 ounces of minced beef
- 1 ½ cloves of garlic, minced.
- 1 teaspoon of Kosher salt
- 1 teaspoon of sesame oil
- 2 eggs
- ½ cup of wheat flour
- 2 tablespoons of grapeseed oil

Dipping Sauce (Vinegary Soy Sauce):
- 2 tablespoons of soy sauce
- ½ teaspoon of rice vinegar
- ¼ teaspoon of roasted and crushed sesame seeds
- ¼ teaspoon of organic sugar

Method:

1. Mince the beef and season it with salt, garlic, and sesame oil. Set aside.

2. Cut off each end of the green peppers, clean out the inside, and wash them with cold running water. Parboil the peppers and then drain. Set them aside.

3. Coat the inner walls of the green peppers with flour, and then pack the seasoned beef (step 1) tightly into the peppers. Lightly coat the packed beef with flour.

4. Crack the eggs and whisk them with a chopstick or fork. Heat a frying pan and add grapeseed oil. When the oil is hot, stir-fry only the part of the pepper with the stuffed beef. When the packed beef inside (#3) take on a yellow tint, flip it over and cook for an additional 2-3 seconds.

5. Serve the stuffed green peppers with the dipping sauce.

Batter-fried Stuffed Mushrooms (Beoseot Jeon / 버섯전)

Total Time: 1 hour 30 minutes

Ingredients (for 4 persons):
- 8 medium-sized shiitake mushrooms (1 ounce total)
- 2 ounces of beef, minced
- 2 ounces of firm tofu (⅕ of a 10-ounce box)
- ½ clove of garlic, diced
- ½ teaspoon of Kosher salt
- 1 teaspoon of roasted sesame oil
- pepper, to taste
- 1 large egg
- ¼ cup of wheat flour
- 1 tablespoon of grapeseed oil

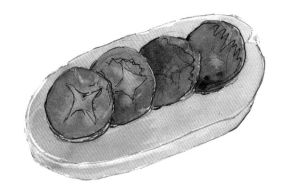

Dipping Sauce (Vinegary Soy Sauce):
- tablespoons of soy sauce
- ½ teaspoon of rice vinegar
- ¼ teaspoon of roasted and crushed sesame seeds
- ¼ teaspoon of organic sugar

Method:

1. Immerse the shiitake mushrooms in water for 1 hour to let them swell. Remove the tough stems, squeeze the mushroom caps to remove excess moisture, and season them with garlic, and sesame oil. Set them aside.

2. Mince the beef. Break up the tofu and remove excess water by squeezing it in a cotton pouch. Mix the minced beef and tofu together, seasoning with salt, pepper, and sesame oil.

3. Sprinkle flour on the inside of each mushroom cap. Fill each cap with the mixture (#2), packing it tightly. Sprinkle flour over the top of the filled mushrooms. Dip each mushroom into the beaten egg, ensuring it's fully coated.

4. Heat the grapeseed oil in a pan over medium heat. Place the stuffed mushrooms in the pan, cooking until they become golden yellow on one side. Flip them over and cook for an additional 2~3 seconds until they are fully cooked.

5. Serve your stuffed mushrooms with the dipping sauce. These batter-fried stuffed mushrooms pair nicely with batter-fried stuffed green peppers as well. Enjoy your meal!

BBQ Beef Short Ribs (LA Galbi / 엘에이갈비)

 Note

With a powerful electric slicer, the butcher cuts the beef ribs very thinly (about ¼ to ⅓ inch thick) across the bone, including three pieces of bone within each slice.

Total Time: 1 hour 30 minutes

Ingredients (for 8 persons):
- 5 pounds of beef short ribs
- 6 tablespoons of organic sugar (for youth, use 8 tablespoons)
- 10 tablespoons of soy sauce (for youth, use 8 tablespoons)
- 1 medium-sized sweet onion
- ½ of large Asian pear
- 4 stalks of scallion
- 6 cloves of garlic
- 2 tablespoons of mirin
- 2 teaspoons of pepper
- crushed roasted sesame seeds.
- roasted sesame oil.

Method:

1. Rinse the beef short ribs under cold water. To remove any bone fragments created during

the slicing procedure, gently spread the meat as if opening a blanket and rub it with both hands under running water to wash away the fragments. Drain them in a large basket.

2. After they are fully drained, sprinkle sugar on each surface of the beef short ribs and wait until all the sugar granules are melted. The sugar serves as a meat tenderizer. Wait for at least 30 minutes at room temperature.

3. In a food processor, grind the onion, garlic, and pear. A small garlic chopper might be too small for this task.

4. Prepare the marinade by combining soy sauce, chopped scallion, the mixture from step 3,

mirin, pepper, and sesame seeds.
5. In a large bowl, place the drained beef short ribs, and then add the mixture from step 4. Mix everything evenly.
6. Add sesame oil last and mix the entire mixture once more. Allow it to marinate for at least 1 hour before serving.

You can choose to cook this dish by BBQ, grilling, or pan-frying it to your preference.

Beef Braised in Soy Sauce (Jangjorim / 장조림)

Total Time: 1 hour

Ingredients (for 4 persons):

- 1 pound of beef, eye of round
  (cut into a 3-inch x 2-inch cuboid)
- ⅓-inch piece of ginger
- 1 tablespoon of mirin
- 2 tablespoons of soy sauce
- 1 tablespoon of soup soy sauce
- ½ of a large apple
- ½ of a large-sized Asian pear
- 6 cloves of garlic
- ½ stalk of big green onion
- 1 stalk of celery
- 1 of a medium-sized onion
- one medium-sized pepper, slightly torn (1 ounce)
- 3 cups of water

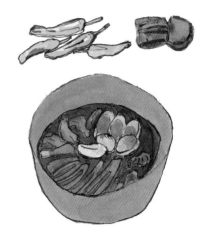

Method:

1. In a 4-quart pot, bring 3 cups of water to a boil. Add the beef, ginger, and mirin. Boil over high heat. Once it starts boiling, skim off any impurities or foam that form on the surface. Boil for about 10 minutes until the center of the beef is fully cooked.

2. Discard half of the boiled water. To the pot, add soy sauce, soup soy sauce, apple, pear, garlic, big green onion, celery, onion, and red pepper. Braise the mixture over medium heat for about 20 minutes.

3. Tear the braised beef into pieces by hand. The pieces can vary in size but should approximately match the thickness of a pencil. You can optionally use a knife instead, but tearing by hand will result in a better texture.

Tips:

- You can also braise boiled eggs, quail eggs, or shishito peppers along with the beef for additional flavors.
- This recipe uses fruit to naturally sweeten the dish, so there's no need to add sugar. The combination of fruits provides a tender and flavorful taste.

Braised Beef Short Ribs (Galbijjim / 갈비찜)

Standard Marinade Ratio for Beef Short Ribs:
Sugar to Soy Sauce Ratio 3.5 : 4.5
For Children & Youth 4 : 4

Total Time: 1 hour

Ingredients (for 8 persons):
- 5 pounds of beef short ribs (Regular short ribs, one piece of bone in one cut, 2~3 inches long)
- 6 tablespoons of organic sugar
- 10 tablespoons of soy sauce
- 4 tablespoons of garlic
- ⅔-inch piece of ginger
- ½ stalk of big green onion
- 1 medium-sized onions
- 1 medium-sized potato
- 1 medium-sized sweet potato
- 2 medium-sized carrots
- 8 dried shiitake mushrooms (1 ounce)
- 10 ginkgo nuts (optional)
- 5 chestnuts (optional)
- 2 tablespoons of mirin
- 1 teaspoon of pepper
- crushed roasted sesame seeds.
- roasted sesame oil.

Method:

1. Fill a large pot (10~12 quarts) with water and bring it to a boil.
2. When the water boils, add the beef short ribs, half of the ginger, and half of the mirin. After about 15 minutes, remove any scum that forms. Wash the short ribs with cold running water and drain them in a strainer. Set them aside.
3. Prepare the marinade by mixing sugar, soy sauce, garlic, big green onion, the remaining half of ginger, the remaining half of mirin, and pepper.
4. Cut the potato, sweet potato, carrot, and onion into chunks.

5. Coat all sides of the drained short ribs with the marinade (#3) and arrange them evenly at the bottom of a deep pot. Place all the vegetables from step 4 on top of the short ribs, with the onion on the very top.

6. Braise the mixture, starting with high heat and then reducing to medium heat. Cook until the surface of the meat becomes slightly firm, and then switch to medium heat. This will help retain the delicious meat juices.

7. Remove excess fat from the surface of the dish. This is an important step for health, so don't skip it. You can do this by refrigerating overnight or using a 5-pound-bag of ice.

 Notes

• Cooking is a kind of science, and special flavors develop as the ingredients work together during the cooking process.

• When purchasing Galbi, you can ask the butcher to trim the white fatty part. Alternatively, you can choose to leave it as is, as it contains nutrients like Glucosamine and Chondroitin. Just remove the excess fat during the skimming process.

Braised Chicken (Dakjjim / 닭찜)

Total Time: 1 hour

Ingredients (for 4 persons):
- 2 pounds of chicken (drumsticks and thighs)
- 2 large potatoes
- 1 large carrot
- 1 large onion
- 1 stick of celery
- ½ stem of big green onion
- 3 cloves of garlic
- ⅓ -inch piece of ginger
- ½ cup of soy sauce
- 2 tablespoons of organic sugar
- ¼ teaspoon of pepper
- 2 tablespoons of mirin
- 1 tablespoon of roasted sesame oil
- ¼ teaspoon of roasted sesame seeds

Method:

1. Begin by cleaning the chicken. You can use a combination of drumsticks and thighs.

2. Parboil the chicken from step 1 with ginger and 1 tablespoon of mirin for about 3 minutes. Then, wash them under cold running water and drain. Set them aside.

3. Cut the potatoes, carrots, onions, big green onions, and celery into large pieces. Set them aside.

4. Create a chicken sauce by mixing diced garlic, soy sauce, sugar, pepper, 1 tablespoon of mirin, and ½ tablespoon of sesame oil.

5. Dip and coat the chicken pieces from step 2 with the chicken sauce you prepared in step 4. Place them at the bottom of the braising pot. Then, simply arrange the vegetables from step 3 on top.

6. Braise the chicken and vegetables: Start on high heat for the first 10 minutes, then reduce to medium heat for the remaining 20 minutes.

7. Finally, add ½ tablespoon of sesame oil before serving.

 Note

The exceptional taste of this dish comes from the harmonious combination of all the ingredients working together during the cooking process.

Braised Croaker (Jogijjim / 조기찜)

Total Time: 1 hour

Ingredients (for 2 persons):

- 2 croakers
  (12 ounces, 15 ounces before cleansing)
- 2 teaspoons of coarse sea salt
- 2 tablespoons of soy sauce
- 1 tablespoon of soup soy sauce
- 1 cup of water
- 4 tablespoons of mirin
- ½ stalk of big green onion with root
- ⅔ -inch piece of ginger
- 1 clove of garlic
- 1 ounce of red pepper (optional)

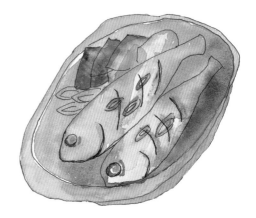

Method:

1. Cleanse the croaker, or you can ask your fish market to do it for you. Try to remove any remaining scales using the back of a big spoon. Brine the surface of the croaker with coarse sea salt, then place it in the freezer.

2. Spread out two pairs of chopsticks on the bottom of a braising pan and place the croaker on top. Mix 1 cup of water with 2 tablespoons of soy sauce, 1 tablespoon of soup soy sauce, and 4 tablespoons of mirin.

3. Pour the mixture from step 2 over the croaker. Add the big green onion, ginger, garlic, and red pepper to one side. Cover with a lid and start broiling over strong heat for about 8 minutes, then reduce the heat to medium and cook for another 12-15 minutes.

4. Uncover the lid and check if most of the liquid has evaporated. If there's still a lot of liquid left, continue to cook until only about ⅛ of the liquid remains.

5. Serve the braised croaker in the braising pan. Dip the fish into the soy sauce liquid beside it.

Braised Fishcake (Eomuk Bokkeum / 어묵볶음)

Total Time: 35 minutes

Ingredients (for 2 persons):
- 5 sheets of rectangular fishcake (7 ounces)
- ⅛ teaspoon of sliced ginger
- ½ sweet onion, chopped
- ½ carrot, chopped
- 2 stalks of scallion, chopped
- 2 cloves of garlic, minced
- roasted sesame seeds
- 3 tablespoons of water
- 1 ½ tablespoons of soy sauce
- 1 teaspoon of mirin
- 1 tablespoon of organic sugar
- 2 teaspoons of roasted sesame oil

Method:

1. Parboil the fishcake in boiling water with a slice of ginger. Then, wash it with cold water. This step helps remove any old frying oil from the fishcake.

2. Heat a frying pan and begin by stir-frying the garlic with oil. Add the parboiled fishcake and water. Once the water is reduced to half, and the fishcake has softened, add the carrot, onion, soy sauce, sugar, mirin, and scallion. Continue stir-frying. Add sesame seeds and sesame oil last.

Braised Half-dried Pollock (Kodari Jjim / 코다리찜)

Total Time: 35 minutes

Ingredients (for 4 persons):

- 2 medium half-dried kodari (3 pounds)
- ¼ of small-sized Korean radish, chunk-cut (¾ cup)
- ⅛ of medium-sized onion, chopped
- 1 tablespoon of red pepper paste
- 2 tablespoons of red pepper powder
- 8 tablespoons of soy sauce
- 1 tablespoon of anchovy sauce
- ¼ stalk of big green onion, cut into 2-inch pieces
- ½ stalk of scallion, chopped
- ½ of red pepper, sliced sideways
- 2 cloves of garlic, minced
- ⅓-inch piece of ginger, sliced
- 2 tablespoons of organic sugar
- ½ teaspoon of roasted sesame seeds
- ½ teaspoon of roasted sesame oil
- 2 cups of water

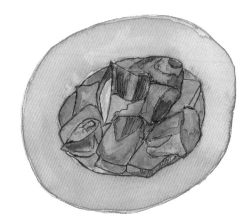

Method:

1. Remove the tail and fins from the half-dried pollock and wash them in cold running water. Ensure that you clean the part where the intestines used to be to avoid a bitter taste.

2. Prepare a marinade by mixing the red pepper paste, red pepper powder, soy sauce, anchovy sauce, sugar, minced garlic, and sliced ginger. Ensure that all the ingredients are well combined.

3. In a wide pot, start by placing the radish at the bottom. Add the pollock, chopped onion, green pepper, and big green onion on top. Pour the marinade (#2) and water over high heat. After boiling for about 2 minutes, cover the pot with a lid and braise for approximately 10 minutes over medium heat. When the liquid has reduced by half, add the sliced red pepper. Finish with roasted sesame seeds and sesame oil. The anchovy sauce adds a savory umami flavor to the dish.

Braised Mackerel (Godeungeo Jorim / 고등어조림)

Total Time: 40 minutes

Ingredients (for 2 persons):
- 1 mackerel (12 ounces)
- 2 tablespoons of soy sauce
- 1 tablespoon of red pepper powder
- 2 cloves of garlic, minced
- 1 teaspoon of minced ginger
- ¼ of small Korean radish, cut into ¾-inch-thick slices.
- ¼ of medium-sized onion, chopped
- ½ stalk of scallion, chopped
- ½ of green pepper, chopped
- 2 teaspoons of mirin
- sprinkle of roasted sesame seeds
- ½ cup of water

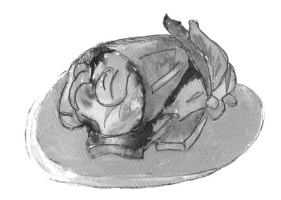

Method:

1. Prepare the marinade by combining soy sauce, red pepper powder, minced garlic, minced ginger, chopped scallion, chopped green pepper, mirin, and sesame seeds.
2. Spread the radish slices on the bottom of a braising pan, then place the mackerel pieces on top of the radish.
3. Evenly pour the marinade (#1) over the mackerel, as if it were raining sauce. Rinse the remaining marinade bowl with water and swirl it around on the bottom of the pan.
4. Start over high heat. Once it begins to boil, reduce the heat to medium and braise for 5-10 minutes. When the marinade liquid has reduced to about 10%, turn off the heat.

You can also use other types of fish for braising, such as hairtail, codfish, haddock, Pacific saury, or Spanish mackerel. If you prefer using red pepper paste, you can replace some of the soy sauce with red pepper paste to suit your taste.

Braised Stalked Sea Squirts (Mideodeok Jjim / 미더덕찜)

Total Time: 30 minutes

Ingredients (for 4 persons)

- 2 cups of stalked sea squirts (cleaned in sea water)
- 3 cups of soybean sprouts (cleaned, washed, and drained)
- ⅓ bunch of dropwort, washed and cut into 1.5-inch pieces
- ¼ of onion, thickly sliced
- ½ stalk of big green onion, chopped
- 1 cup of anchovy broth, chicken broth, or vegetable broth
- ⅛ teaspoon of Kosher salt
- 1 teaspoon of roasted sesame seeds
- 1 ½ teaspoons of roasted sesame oil

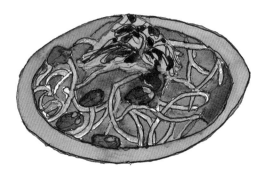

Seasoning Sauce:

- 1 tablespoon of grapeseed oil
- 3 tablespoons of red pepper powder
- 3 cloves of garlic, minced.
- 2 tablespoons of soy sauce
- 1 tablespoon of mirin or white wine
- pepper (to taste)

Starch Water:

- 3 tablespoons of corn starch

- ¾ cup of water

Method:

1. In a heated pot, stir fry the ingredients of the seasoning sauce. Then, add soybean sprouts, cover with a lid, and cook for 2 minutes over high heat. Add the stalked sea squirts, anchovy broth, onion, big green onion, and continue boiling for 2 more minutes.

2. Reduce the heat and pour in the starch mixture (corn starch water) over the contents of the pot. Stir and simmer for 1 additional minute. Add sesame seeds and dropwort, then swirl in the roasted sesame oil. Serve.

You can enhance the dish by adding small shrimps, mussels, cuttlefish, or angler pieces.

Angler Variation
(Braised Soybean Sprout & Angler):
1. Chop an angler into small pieces. Blanch the angler pieces until the skin turns white. Wash them with cold water and drain.
2. Stir fry the angler meat and liver in some oil until fully cooked, about 8 minutes. Then, proceed with step 1 of the recipe "Braised Soybean Sprout & Stalked Sea Squirts."

Broiled Pork Ribs (Dwaeji Galbi / 돼지갈비)

Total Time: 1 hour 30 minutes

Ingredients (for 8 persons):

- 5 pounds of pork ribs
- 1 tablespoon of soybean paste
- 1 tablespoon of instant coffee
- 2-inch piece of ginger
- 6 quarts (1 ½ gallons or 24 cups) of water
- 5 tablespoons of soy sauce
- 3 tablespoons of red pepper paste
- 1 cup of organic sugar2 tablespoons of red pepper powder
- 1 medium-sized sweet onion
- 4 stalks of scallion
- 3 cloves of garlic
- 2 tablespoons of mirin
- 1 teaspoon of pepper
- 1 tablespoon of crushed roasted sesame seeds
- 1 tablespoon of roasted sesame oil

Method:

1. Fill a large pot (10-12 quarts) about ⅔ full of water.

2. Add soybean paste, instant coffee, and ginger to the water, and bring it to a boil.

3. Once the mixture in step 2 starts boiling, add the pork ribs and boil for approximately 15 minutes.

4. Remove the pork ribs from the boiling water and rinse them under cold running water. Drain them using a strainer. Set them aside. The right time to stop boiling is when about ⅓ inch of the bone is exposed, and there is no more blood oozing from the bone marrow.

5. In a garlic chopper, grind onion, garlic, and ginger. Set this mixture aside.

6. Create a marinade for the pork ribs by mixing soy sauce, red pepper paste, sugar, corn syrup, red pepper powder, scallion, mirin, pepper, sesame seeds, sesame oil, and the mixture from

step 5.

7. Coat and marinate the pork ribs thoroughly. Spread them evenly on a broiler pan, creating a single layer.

8. Broil the ribs for 20 minutes or until the top part turns light brown. Flip them over and broil for an additional 10 minutes. Alternatively, you can also prepare this dish on a BBQ grill.

Bulgogi (불고기)

Total Time: 1 hour

Ingredients (for 4 persons):

- 1 ½ pounds of shredded beef
  (sirloin, round, plate, flank, any part is okay)
- 3 tablespoons of organic sugar
- 5 tablespoon of soy sauce
- ½ of medium-sized sweet onion
- 3 cloves of garlic
- 4 stalks of scallion
- ¼ of a large Asian pear
- 1 teaspoon of pepper
- 2 tablespoons of mirin
- 1 tablespoon of roasted and crushed sesame
  seeds
- 2 tablespoons of roasted sesame oil

Method:

1. Coat each piece of shredded beef, front and back, with sugar. Wait until the sugar granules are completely melted. Patience is key; rushing this step can result in stiff and low-quality Bulgogi. Sugar requires time to act as a meat tenderizer. Set it aside.

2. While waiting, grind onion, garlic, and pear together in a small garlic chopper. Set it aside.

3. Combine soy sauce, scallion, pepper, mirin, sesame seeds, and the mixture from step 2. Set it aside.

4. Mix the beef from step 1 and the sauce from step 3. After achieving an even mix, add sesame oil last and mix again.

 Note

You can add shiitake mushrooms, white mushrooms, carrot, or glass noodles according to your taste, but omitting them will preserve the authentic taste of Bulgogi. Mirin can be replaced with other types of liquor used in your cooking. Mirin is a type of rice wine like sake but with lower alcohol content and higher sugar content, which occurs naturally during fermentation with no added sugars.

Cabbage Rolls (Yangbaechu Roll / 양배추롤)

Total Time: 1 hour

Ingredients (for 3 persons):
- 6 wide sheets of cabbage leaves (approximately 8 ounces)
- 5 ounces of ground beef (80% lean)
- 5 ounces of ground pork (80% lean)
- ¼ cup of finely chopped sweet onion
- 3 cloves of garlic, minced
- 1 teaspoon of ginger juice
- 2 teaspoons of mirin
- 1 tablespoon of Kosher salt
- sprinkle of pepper
- 2 eggs
- 2 tablespoons of unsalted butter
- ½ cup of tomato sauce

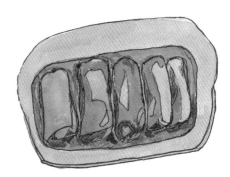

Method:

1. Begin by blanching the cabbage leaves in boiling water with a pinch of salt. After blanching, rinse them in cold water and let them drain. Press down on the thick stalks gently using the back of a knife or a pushing rod.

2. In a large mixing bowl, combine the ground beef, ground pork, chopped sweet onion, minced garlic, ginger juice, mirin, Kosher salt, and a sprinkle of pepper. Add the eggs and mix everything together thoroughly.

3. Take 3-4 tablespoons of the mixture from step 2 and roll it into a neat cuboid shape within each cabbage leaf, ensuring that the edges are tucked in. You should end up with six rolls in total. Place these rolls in a braising pan and pour the tomato sauce over them. I recommend using a pan with a glass lid.

4. Place the braising pan with the rolls over high heat. Once the pan is hot, melt the butter around the rolls, cover the pan with the lid, and bring it to a boil. Once it starts boiling, reduce the heat to medium-low and let it simmer for about 20 minutes.

5. These rolls can be served either hot or cold and paired wonderfully with steamed rice.

Fried Cuttlefish (Ojingeotwigim / 오징어튀김)

Total Time: 1 hour

Ingredients (for 4 persons):
- 2 small cuttlefish (1 pound)
- ⅓ canned pineapple (approximately 5 ounces)
- ⅜ cup of mayonnaise
- ¼ cup of organic sugar
- 2 cups of grapeseed oil (for frying)

For the batter:
- 2 egg white
- 3 tablespoons of wheat flour
- 3 tablespoons of cornstarch

Method:

1. Begin by preparing the cuttlefish. Remove the bone, intestines, peel off the skin, and wash thoroughly. Drain the cuttlefish on a basket.

2. Cut the cuttlefish in half lengthwise, remove the outer sheath, and then slice them into ⅔ -inch width strips.

3. To reduce splattering when frying the cuttlefish, blanch the cuttlefish strips (#2) and drain the excess water.

4. In a bowl, separate the egg whites from 2 eggs and whisk them together. Mix well with the wheat flour and corn starch. The batter may appear thin.

5. Coat the blanched cuttlefish strips (#3) with the egg white batter (#4).

6. In a deep-frying pan, heat the grapeseed oil. Fry the cuttlefish strips in hot oil over medium heat to maintain the whiteness of the batter. Drain excess frying oil and set aside.

7. Take the pineapple chunks from the can and drain any juice from them. Set the pineapple chunks aside.

8. In another frying pan, mix the pineapple

chunks (#7), organic sugar, and mayonnaise. Stir quickly over medium heat until they achieve a suitable transparent tint. Pour this mixture over the fried cuttlefish (#6) and serve.

Tip:

You can experiment with different fruits like kiwi or orange pieces instead of pineapple, as suggested. It can provide a unique and delicious twist to the dish.

Four Different Ways to Make Frying Batter (Coating):

1. Using watery starch for a very thick coating, as seen in Sweet and Sour Pineapple Pork.
2. Using egg whites along with flour and starch for a dilute batter, as used in Fried Cuttlefish.
3. Using egg yolks along with corn starch powder for a soft-textured batter, as in Lemon Chicken.
4. Using a very small amount of dry corn starch to add a thin layer, as in Fried Chicken.

Hamburger Patties (햄버거패티스)

Total Time: 1 hour

Ingredients (for 4~6 persons):
- 2 pounds of ground beef (90% lean)
- 1 egg, beaten
- ¾ cup of dry breadcrumbs
- 3 tablespoons of evaporated milk
- 2 tablespoons of Worcestershire sauce
- 2 cloves of garlic, minced
- ⅛ teaspoon of cayenne pepper (optional)

Method:

1. In a large bowl, combine all the ingredients: ground beef, beaten egg, dry breadcrumbs, evaporated milk, Worcestershire sauce, minced garlic, and cayenne pepper (if using). Mix everything together well. Be careful not to over-knead the mixture; you want to maintain the meat's texture. If needed, you can add more dry breadcrumbs to help shape the meat into patties.

2. Shape the mixture into patties that are about ½-inch thick. Cover the patties with plastic wrap and refrigerate them for about 30 minutes. This helps the patties hold their shape while cooking.

Baking Method:

1. Preheat your oven to 420 degrees Fahrenheit.

2. Heat a frying pan and briefly stir-fry both sides of the patties until just the surfaces are cooked.

3. Place the partially cooked patties on a baking sheet and bake them in the preheated oven. Bake one side for 15 minutes, then flip the patties and bake the other side for another 15 minutes. Baking will result in softer and juicier patties.

Grilling Method:

1. Prepare your grill with a strong flame or grill setting.

2. Coat the grill grid with oil to prevent sticking. Grill one side of the patties for about 5 minutes, then flip them and grill the other side for another 5 minutes. Keeping the meat juicy over strong heat at the beginning of cooking is crucial for a delicious result.

Enjoy your homemade beef patties, whether baked or grilled!

Hand Rubbed Beef (Jumulleok / 주물럭)

Total Time: 1 hour

Ingredients (for 4 persons):
- 2 pounds of flap meat
- 5 tablespoons of organic sugar
- 5 tablespoons of soy sauce
- ⅓ stalk of big green onion
- 1 medium-sized sweet onion
- 3 cloves of garlic
- 1 tablespoon of mirin
- 1 teaspoons of pepper
- 1 tablespoon of sesame seed roasted & crushed
- 1 tablespoon of sesame oil, roasted

Method:

1. Lay down flap meat on a cutting board and cut the meat vertically (a little sideway for your knife to face meat maxim width).

2. Sprinkle sugar evenly and coat # 1 meat until sugar melts transparently. Set aside. You never hurry. Be patient; remember to give sugar time to accomplish its task as a meat tenderizer.

3. Make marinade with soy sauce, grated garlic and onion, and sesame seed.

4. Mix #2 meat with #3 mixture. Add sesame oil last. Set it aside.

5. BBQ or pan fry.

Let's see the difference between Bulgogi and Jumulluk

Bulgogi 2 LB = 6 LBs sugar / 6 LBs soy sauce
Jumulluk 2 LB = 5 TBs sugar / 5 LBs soy sauce

King Dumpling (Wang Mandu / 왕만두)

Total Time: 2 hours

Ingredients:
For the Dumpling Dough:

- 3 cups of wheat flour, all-purpose
- 2 teaspoons of raw yeast
- 1 ⅓ cups of warm water

Method:

For the Dumpling Dough:

1. In a small bowl, dissolve the raw yeast in warm water and mix well.

2. In a large bowl, place the all-purpose flour. Gradually add the raw yeast water to the flour while mixing with your hand until it forms a cohesive dough.

3. Once the dough comes together, transfer it to a cutting board. To develop gluten and create a smoother texture, repeatedly throw the dough onto the board, and hold it in your hand for at least 30 minutes.

4. Place the dough in a warm location, covered with a damp cotton cloth, until it has expanded to 1.5 times its original size. Repeat the "throw and hold" process one more time. This is the first fermentation.

5. Roll out the dough into round, flat dumpling

wrappers, each about 4 inches in diameter.

For the Filling:

Please see the separate section below for the Dumpling (Meat & Vegetable) Filling.

1. Place a portion of the filling in the center of each dumpling wrapper. Fold the wrapper over the filling to create a half-moon shape. Press and twist the center to seal the dumpling closed securely.

2. Line a bamboo steamer with a damp burlap cloth. Place the king dumplings on the cloth, making sure they are spaced apart, and allow them to ferment for an additional hour. This is the second fermentation.

3. Prepare a pot of boiling water and place the bamboo steamer over it. Cover the steamer with its bamboo lid and steam the king dumplings for about 20 minutes. Do not open the lid during this time.

These King Dumplings are a delightful treat, and you can choose to fill them either with meat and vegetables or sweet red bean paste. Enjoy!

Dumpling (Meat & Vegetable) Filling:

Ingredients:
- ¼ cup of ground beef (80% lean)
- ¼ cup of ground pork (80% lean)
- ⅕ bundle of chives (½ bundle) (½ cup)
- 2 leaves of Napa cabbage (⅛ cup)
- 1 package of mung bean sprouts (3.7 ounces)
- ¼ medium zucchini
- half of a medium-sized onion
- 2 cloves of garlic
- ⅔ -inch piece of ginger
- ⅛ stalk of big green onion
- 1 teaspoon of mirin
- ½ box of tofu with the least amount of water inside (5 ounces)
- 1 egg
- ½ teaspoon of Kosher salt
- 1 teaspoon of soy sauce
- sprinkle of pepper
- 1 teaspoon of roasted sesame oil

Method:
1. Blanch the Napa cabbage and mung bean sprouts, chop them into small pieces, and squeeze out excess moisture from each.
2. Slice the zucchini into 1-inch slices, sprinkle with salt, and squeeze to remove excess moisture.
3. Slice and dice the onion, then squeeze out excess moisture.
4. Place the tofu in a burlap pouch and squeeze out excess water.
5. In a big bowl, add the ingredients from steps 1 to 4, along with the ground beef, ground pork, salt, soy sauce, pepper, and sesame oil. Mix everything well.
6. Add the minced garlic, ginger juice, big green onion slices, and mirin to the mixture. Mix thoroughly.
7. Lastly, add the beaten egg to the mixture and mix until all the ingredients are well combined.

This filling is ready to be used for making your King Dumplings. Enjoy your dumplings with this delicious filling!

Korean Patties (Wanjajeon / 완자전)

Total Time: 1 hour

Ingredients (for 4 persons):
- 1 pound of ground beef (90% lean)
- 1-inch piece of ginger, minced
- 1 tablespoon of diced scallion
- 3 cloves of garlic, minced
- ½ tablespoon of soy sauce
- ½ teaspoon of Kosher salt
- ¼ teaspoon of pepper
- 1 tablespoon of cornstarch
  (or more, as needed)
- ¼ tablespoon of roasted sesame oil
- 1 egg
- ½ tablespoon of mirin

For Coating:
- ¼ cup grapeseed oil (for stir-frying)
- ¼ cup of wheat flour (for coating)
- 3 eggs (for coating)

Method:

1. In a big bowl, vigorously mix the ground pork, minced ginger, diced scallion, minced garlic, soy sauce, Kosher salt, pepper, roasted sesame oil, egg, and mirin. Adjust the amount of cornstarch as needed. The goal is to create a mixture that resembles dough. Stir vigorously with chopsticks

and knead with your hands if necessary.

2. Shape the mixture into patties with a 1.5-inch diameter. You should be able to make around 20 patties.

3. In a small bowl, whisk the eggs, and spread the wheat flour on a dish.

4. Coat each patty with flour, dip it into the beaten egg, and then stir-fry it in grapeseed oil over high heat. Once the patties have a golden-brown tint on their surfaces, reduce the heat to medium-low. Occasionally, wipe away any browned residue from the pan with a paper towel.

5. Serve the pork patties with a dipping sauce made from vinegar and soy sauce in a 1:1 ratio.

Korean Style Egg Roll (Gyeran Mari / 계란말이)

Total Time: 30 minutes

Ingredients (for 2 persons):
- 3 eggs
- ½ teaspoon of Kosher salt
- 1 stalk of scallion
- ¼ of medium carrot
- 1 tablespoon of grapeseed oil
- 2 tablespoons of water (optional)

Method:

1. Dice the scallion and carrot into small pieces. Finely chopped pieces work best for flavor.

2. In a mixing bowl, break the eggs and add the diced scallions, carrots, and salt. Mix the ingredients together. You can add water to the mixture if desired, but it's optional.

3. Prepare an 8-inch square or rectangular-shaped frying pan to make even egg rolls.

4. Heat the pan over high heat, swirl around the grapeseed oil, and remove any excess oil with a paper towel. Then, reduce the heat to low.

5. Pour one-third of the egg mixture (#2) into the pan, ensuring it spreads evenly to cover the pan's surface. Roll it up like a cake roll. When the outside of the egg mixture turns a golden-brown color, push it to one side of the pan. Add another one-third of the egg mixture to the end of the previously cooked eggs. Repeat this process until you've used all the egg mixture. To shape the roll firmly, stand it up on all four sides, pressing the sides with two spatulas each time.

Lasagna

Total Time: 2 hours

Ingredients (for 8 persons):
- 4 cups of lasagna noodles, cooked. (8 ounces)
- 1 ½ cups of ricotta cheese
- ¼ cup of Parmesan cheese
- 1 egg, beaten.
- ⅛ tsp of Kosher salt
- ⅛ tsp of pepper
- 2 cups of Sicilian tomato sauce
- ½ cup of aged mozzarella cheese
- ½ cup of provolone cheese
- grated Parmesan cheese (additional for topping)

Method:

1. Prepare the Ricotta Cheese Mixture:
- In a mixing bowl, combine the ricotta cheese, ¼ cup Parmesan cheese, beaten egg, Kosher salt, and pepper.
- Mix thoroughly to create the ricotta cheese mixture.

2. Prepare the Baking Pan:
- Preheat your oven to 375 degrees Fahrenheit.
- Take a 13"x9"x2" baking pan and lightly grease the bottom and all four sides with oil or cooking spray.

3. Layering the Lasagna:
- First Layer: Spread ½ cup of Sicilian tomato sauce evenly across the bottom of the prepared baking pan.
- Place ⅓ of the cooked lasagna noodles over the sauce.
- Spread ⅓ of the ricotta cheese mixture evenly over the noodles.
- Sprinkle ⅓ of the mozzarella cheese and ⅓ of the provolone cheese over the ricotta mixture.

4. Repeat the Layers:
- Repeat the layering process with the remaining lasagna noodles, ricotta cheese mixture, mozzarella cheese, and provolone cheese.

5. Final Layer:
- For the final layer, pour the remaining Sicilian tomato sauce evenly over the lasagna.
- Add the last ⅓ of the lasagna noodles.
- Spread the last ⅓ of the ricotta cheese mixture.
- Sprinkle the remaining mozzarella cheese and provolone cheese.
- Top it off with additional grated Parmesan cheese.

6. Baking:
- Cover the baking pan with aluminum foil and bake in the preheated oven for approximately 40 minutes or until the lasagna is hot and bubbly.

7. Serving:
- Allow the lasagna to cool for a few minutes before serving
- Slice and serve. It's best when served hot.

This classic lasagna recipe is rich with layers of pasta, cheese, and savory Sicilian tomato sauce. Enjoy your meal!

Mapo Tofu (Mapa Dubu / 마파두부)

Total Time: 30 minutes

Ingredients (for 4 persons):
- 1 box of firm tofu, cut into ½-inch cubes
- 2 ounces of lean ground pork (80% lean)
- 1 tablespoon of diced big green onion
- 3 cloves garlic, diced
- 1 tablespoon of diced ginger
- 1 red pepper diced
- 1 ½ tablespoon of grapeseed oil (for stir-frying)
- 2 cups of canola oil (for deep frying)
- ½ cup of water
- 1 teaspoon of mastic-leaf prickly ash or pepper

Mapo Tofu Sauce:
- 1 tablespoon of chili bean sauce
- 2 tablespoon of soy sauce
- 2 tablespoons of mirin
- ½ tablespoon of organic sugar
- 1 tablespoon of watery starch
- 1 teaspoon of roasted sesame oil

Method:

1. Cut the tofu and dice the big green onion, garlic, ginger, red pepper, and set aside with the

ground pork.

2. Make the Mapo Tofu Sauce by mixing all six sauce ingredients evenly in a bowl.

3. Prepare a deep, wide pan for frying and pour in the canola oil. Heat it. Place tofu cubes in a sieve with a long handle and dip the sieve in and out of the hot oil until the tofu turns yellow. Drain the oil from the tofu cubes and set them aside.

4. Heat a wok and add the grapeseed oil. Once heated, stir-fry the ginger, garlic, red pepper, and big green onion to develop flavor. Then add the ground pork and continue stir-frying.

5. When the pork is cooked, add ½ cup of water.

When the water boils, add the Mapo Tofu Sauce
(step 2) and quickly mix it with the tofu. Sprinkle
with mastic-leaf prickly ash or pepper.

 Note

To make watery starch, mix cornstarch and
water in a bowl. Stir until well combined. Allow
it to settle for several hours. Once the sediment
settles at the bottom of the bowl, pour out the
top water, leaving only the sediments. This
is watery starch, which can be stored in the
refrigerator for future use.

Meat Sauce for Spaghetti in Crock Pot

Total Time: 30 minutes for seasoning and 8 hours in crock pot for cooking

Ingredients (for 8 persons):
- 2 pounds of chopped beef
- 1 pack of Italian sausage (mild, 1 pound)
- 2 medium-sized sweet onions, diced
- 2 cloves of garlic, minced
- ¼ cup of olive oil
- 1 basil stem (2 teaspoons)
- 1 can (28 ounces of Italian-style broken-up tomatoes)
- 1 can (28 ounces of tomato sauce)
- 1 can (5 ounces of tomato paste)

Method:

1. Use a crockpot for this recipe. There are various sizes available; choose one according to your needs. An 8-quart crock pot is recommended for the given amount, but you can scale it down if your family is smaller.

2. Brown the chopped beef and Italian sausage, then add them to the crockpot.

3. In a separate pan, sauté the diced onion and minced garlic in olive oil until they are fragrant and softened. Add this mixture to the crockpot.

4. Add all the other ingredients (basil, Italian-style broken-up tomatoes, tomato sauce, and tomato paste) into the crockpot, and cover it with the lid.

5. Start cooking on low for 8 hours. Use defrosted meat (beef and sausage).

Cooking the Spaghetti:

- Use about 5-6 oz of spaghetti per person. To gauge the amount, hold the raw spaghetti noodles loosely between your thumb and pointer finger; this is roughly the amount for one person.

- Boil the spaghetti noodles for about 8-12 minutes or according to your taste. Drain the hot water but do not rinse the noodles.

- Pour 7-8 oz of the meat sauce you prepared earlier over the cooked spaghetti noodles. Sprinkle some Parmesan cheese before eating to enhance the flavor.

Mung Bean Pancake (Bindaetteok / 빈대떡)

Total Time: 2 hours

Ingredients (for 4 persons)
- 2 cups of mung beans
- 3 tablespoons of swollen rice
- 3 ounces of pork
- ⅓ cup of mung bean sprouts (1 ⅕ ounces)
- fistful of dry bracken= ⅔ ounces
- (⅓ cup of swollen bracken)
- ¼ cup of swollen bellflower
- ¼ stalk of big green onion
- 4 leaves of Napa cabbage (or kimchi)
- 1 red pepper (1 ounce)
- ⅓ cup of grapeseed oil
- 2 cloves of garlic
- ¼ teaspoon of ginger
- 2 teaspoons of Kosher salt
- 2 teaspoons of roasted and crushed sesame seeds
- 2 teaspoons of roasted sesame oil
- 1 tablespoon of mirin

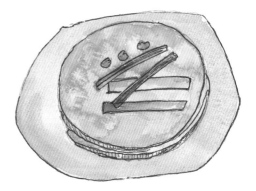

Method:

1. Wash and clean the mung beans. Soak them in lukewarm water for 3-4 hours, then rub and peel the shells from the mung beans. Drain.

2. Using a food processor, grind the soaked and drained mung beans along with the swollen rice.

3. In a chopper, chop the pork and season it with garlic, big green onion, pepper, and ginger.

4. Boil the bracken, drain it. Bathe the bellflower in salt to remove the pungent taste, then cook it in boiling water until softened, and squeeze out the water. Parboil the mung bean sprouts, then drain. Season these three vegetables with garlic, salt, sesame seeds, and sesame oil.

5. Brine the Napa cabbage, wash it, tear it lengthwise, and squeeze out excess water. Slice the big green onion.

6. In a large bowl, mix the ground mung beans

and rice (#2), seasoned pork (#3), seasoned bracken, bellflower, and mung bean sprouts (#4), and brined Napa cabbage and sliced big green onion (#5).

7. Make thick mung bean patties by pan-frying them in heated oil. Decorate the tops of the patties with slices of red pepper and the leafy part of big green onions when you flip them over.

Enjoy your delicious mung bean pancakes = **Bindaetteok!**

Pan-fried Fish Fillet (Saengseonjeon / 생선전)

Total Time: 30 minutes

Ingredients (for 4 persons):
- 1 fish - fresh filet of cod, flounder, or pollock (8 ounces)
- 1 teaspoon of Kosher salt
- ⅛ teaspoon of pepper
- ¼ cup of wheat flour
- 2 eggs
- 2 tablespoons of grapeseed oil

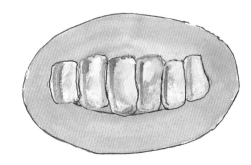

Method:

1. Begin by sprinkling salt and pepper on both sides of the fish filets.

2. Dredge all the fish filets in flour, ensuring an even coating, and then spread them on a wide plate.

3. Beat the eggs in a separate container.

4. Heat a pan over strong heat, then pour in the oil. Once the oil is heated, reduce the heat to medium. Dip each filet from step 2 into the beaten eggs, then place it in the pan. Fry the filet, flip it over as needed, until all filets are cooked.

5. Drain any excess frying oil.

6. Serve and enjoy your delicious fried fish filets.

Tips:

When purchasing fresh fish from the market, ask them to provide 2 whole filets. Upon arriving home, freeze the fish right away. After it is halfway frozen, making thin filets becomes much easier.

Pan-fried Slice of Beef (Yukjeon / 육전)

Total Time: 30 minutes

Ingredients (for 4 persons):
- 4 ounces of beef (rump meat), shredded.
- 1 teaspoon of mirin
- 1 egg
- 2 tablespoons of wheat flour
- ⅛ teaspoon of Kosher salt
- sprinkle of pepper
- 1 tablespoon of grapeseed oil

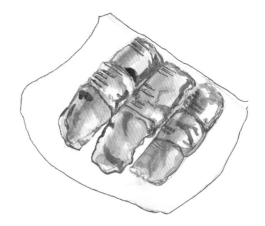

Method:

1. Shred the beef, like how you would prepare it for bulgogi, or you can ask your butcher to shred it like bacon.

2. Score the beef slices across the grain and sprinkle them with salt and pepper.

3. Open and break the egg, then add salt to it.

4. Cover the beef with flour, dip it in the egg, and fry it in a pan over medium heat.

5. Serve the pan-fried beef slices with meat soy sauce.

Optional: Mustard sauce pairs well with pan-fried beef slices.

Meat Soy Sauce:
- 1 tablespoon of soy sauce
- 2 teaspoons of rice vinegar
- ½ + ½ teaspoon of organic sugar

Enjoy your delicious, pan-fried beef.

Passover Broiled Beef Flank

Total Time: 6~8 hours for marinating and 20 minutes for cooking

Ingredients (for 4 persons):
- 1 pound of beef flank (1 package)
- 1 medium-sized onion
- 2 cloves of garlic
- 2 tablespoons of soy sauce
- 2 tablespoons of honey
- 2 tablespoons of water

Method:

1. Place the beef flank in a one-gallon Ziploc bag. Cut the medium-sized onion in half crosswise and use the open side of the onion to rub vigorously on each surface of the beef flank until the onion turns a deep red color.
2. Remove the red-stained onion parts and chop the remaining onion. Slice the garlic. Divide the chopped onion and sliced garlic between both sides of the beef flank. Pour one tablespoon each of soy sauce, honey, and water onto each side as well. Seal the Ziploc bag and marinate in the refrigerator for at least 6 to 8 hours.
3. Preheat your grill or broiler. Grill the marinated beef flank for about 6 to 7 minutes on one side, then flip it over and grill for an additional 3 to 4 minutes.
4. Allow the grilled beef flank to cool and then refrigerate it. After one or two days, slice it very thinly before serving.

Platter of Nine Delicacies (Gujeolpan / 구절판)

Total Time: 2 hours

Preparing Eight Delicacies

Ingredients (4 persons):
- 4 ounces of lean beef, sliced into fine strips along the grain
- 4 ounces of raw shiitake mushrooms, sliced into strips
- 1 medium carrot, sliced into strips
- 1 cucumber, sliced into strips
- ¼ of small Korean radish, dyed in beet juice and sliced into strips
- 2 green peppers, sliced into strips (2 ounces)
- 2 eggs, pan-fried and cut into white and yellow strips
- 1 tablespoon of Kosher salt
- 2 tablespoons of grapeseed oil

Method:

1. Slice the beef into fine strips along the grain. Season it with soy sauce, pepper, sugar, garlic, and mirin. Stir-fry the beef in a pan with a minimal amount of oil. Let it cool and set it aside.

2. Slice the mushrooms into thin strips and season them with salt and pepper. Set them aside.

3. Remove the seeds from the cucumber and slice it into fine strips. Sprinkle salt on them, and once they become limp, stir-fry them in a pan with a little oil. Do the same for the carrots. Set both aside.

4. Dye the radish strips with beet juice for 5 minutes to give them a purple tint. Sprinkle salt on them, and once they become limp, stir-fry them in a pan with a minimal amount of oil. Set them aside.

5. Cut the green peppers into thin strips, stir-fry them in a pan with a minimal amount of oil, and let them cool. Set them aside.

6. Break the eggs and separate the egg yolks and whites. Stir-fry them in a pan with a minimal amount of oil to make thin layers of each. Cut these layers into fine strips of the same length as all the vegetables. Set both aside.

7. Arrange each of the eight delicacies, prepared through the procedures from steps 1-6, by different colors in one of the eight side compartments of a nine-platter dish. Set it aside.

Making the Crepes for the Center of the Platter

Ingredients:

- 1 cup of wheat flour
- 1 cup of water
- 1 teaspoon of Kosher salt
- 1 tablespoon of grapeseed oil
- 2 tablespoons of pine nut powder

1. Mix flour, water, and salt well. Let the mixture stand for 1 hour, then pass it through a sieve.
2. Heat a frying pan, swirl some oil, wipe it with a paper towel, and reduce the heat to low. Drop a spoonful of batter to make very thin crepes, each with a diameter of 3.5 inches. Keep making crepes until no batter remains. Sprinkle pine nut powder on each crepe to prevent sticking.

Your eight delicacies and crepes are now ready to be enjoyed as a colorful and flavorful dish!

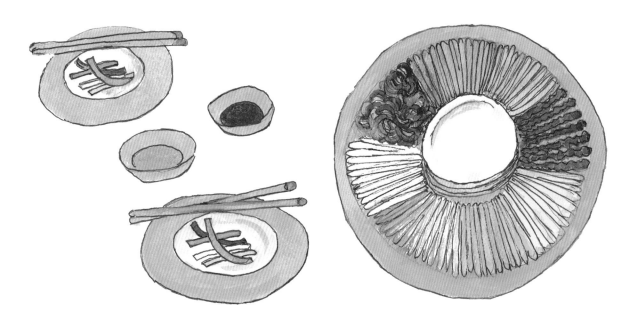

Pork Bulgogi (Dwaeji Bulgogi / 돼지불고기)

Total Time: 1 hour

Ingredients (for 4 persons):
- 1 ½ pounds of shredded pork (round or butt, any part fine)
- 4 ½ tablespoons of organic sugar
- 4 ½ tablespoons of soy sauce
- 2 tablespoons of red pepper powder
- ½ sweet onion, sliced
- 3 cloves of garlic, minced
- 1-inch piece of ginger, minced
- 4 stalks of scallion, chopped
- ½ green pepper, chopped sideways (½ ounce)
- ½ red pepper, chopped sideways (½ ounce)
- ¼ of a large Asian pear
- 1 teaspoon of pepper
- 2 tablespoons of mirin
- 1 tablespoon of roasted and crushed sesame seeds
- 1 tablespoon of roasted sesame oil

Method:

1. Coat each piece of shredded pork, front and back, with organic sugar. Allow the sugar granules to melt transparently. Be patient as this step is crucial for tenderizing the meat. Set it aside. Bulgogi Standard Ratio: For every 1 pound of pork, use 3 tablespoons of soy sauce and 3 tablespoons of sugar.

2. While waiting, slice the onion, chop the green pepper and red pepper, and grind the garlic, ginger, and pear together in a garlic chopper. Set it aside.

3. In a separate bowl, combine soy sauce, scallions, red pepper powder, pepper, mirin, and sesame seeds with the mixture from step 2. Set it aside.

4. Mix the sugar-coated shredded pork from step 1 with the mixture from step 3. Once everything is evenly mixed, add the sesame oil as the final step.

You can replace mirin with another type of liquor that you prefer for your cooking.

Enjoy your delicious, shredded pork bulgogi!

Seafood Dynamite

Total Time: 40 minutes

Ingredients (for 4 persons):

- ½ sweet onion, diced
- 10 white button mushrooms, diced (3 ounces)
- ¼ cup of shelled shrimp (2 ounces)
- ¼ cup of bay scallops (2 ounces)
- ¼ cup of crab meat (2 ounces)
- ¼ cup of shelled clam meat (2 ounces)

Dynamite Sauce:

- 6 tablespoons of mayonnaise
- ½ teaspoon of lime juice
- 1 teaspoon of organic sugar
- 1 tablespoon of smelts' roe

Method:

1. Heat up the oven to 400 Degrees Fahrenheit.
2. Mix dynamite sauce well. The pink color looks pretty.
3. Pat and try to dry 4 kinds of seafood with paper towels.
4. Stir and mix half of the dynamite sauce with seafood, onion, and white button mushrooms.

5. Prepare oven-safe container, spread out # 4 mixture, cover the top with the remaining dynamite sauce.
6. Bake for 20 minutes until the seafood dynamite turns slightly brown. Dish out and serve hot.

Seafood Scallion Pancake (Pajeon / 파전)

Total Time: 30 minutes

Ingredients (for 4 persons):

- 1 cup of seafood (shrimps, calamari, or clams/oysters) (8 ounces)
- 1 cup of sifted all-purpose flour
- 1 tablespoon of rice powder
- 1 cup of icy cold water
- ½ teaspoon of Kosher salt
- 1 egg, beaten
- 1 tablespoon of roasted sesame oil
- 2 bunches of scallions, use the green part
- sprinkle of pepper
- ⅜ cup of grapeseed oil

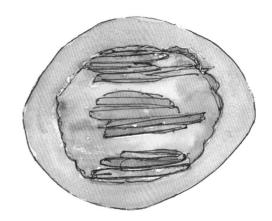

Method:

1. Prepare the batter by mixing flour, rice powder, salt, and egg with icy cold water.
2. Preheat your frying pan, add grapeseed oil, and evenly spread the oil around the pan.
3. Once the oil is sufficiently heated, pour in half of the batter and spread it evenly in a circular shape.
4. Place half of the green scallion on top of the batter, followed by the shrimp and calamari (or other seafood) on top of the scallion. Repeat with the remaining seafood. Gently press down with a spatula.
5. When the bottom part is golden brown and has hardened, flip it over. You'll end up with two pieces of pajeon using this method.

Tips:

- Avoid using mixed products with taste enhancers and preservatives for a healthier and crispier pajeon.
- You only need to flip it once, like making a pancake.

Seasoned Dried Pollock (Bugeopo Muchim / 북어포무침)

Total Time: 30 minutes

Ingredients (for 2 persons):
- 1 small package of torn dried pollock (3 ½ ounces)
- ½ of Asian pear, grated and sieved
- 3 cloves of garlic, minced
- 1 tablespoon of red pepper powder
- 2 tablespoons of red pepper paste
- ¼ white stalk of big green onion
- 1 tablespoon of honey
- 1 tablespoon of soup soy sauce
- 1 teaspoon of roasted sesame seeds
- 1 teaspoon of roasted sesame oil

Method:

1. Tear the whole dried pollock by hand the previous night.

2. Grate ½ a pear using a grater, pass it through a sieve, and squeeze out all the juice. Soak the torn dried pollock in the pear juice overnight. The tough dried pollock softens in the pear juice.

3. In a large bowl, place the softened pollock from step 2. Mix it well with garlic and red pepper powder. Continue mixing by adding red pepper paste, big green onion, honey, and soup soy sauce. Cover with a lid and refrigerate. When you're ready to eat, mix it with sesame oil, and sprinkle sesame seeds on top. Serve and enjoy.

Seasoned Shredded Squid (Ojingeo Chae Bokkeum / 오징어채볶음)

Total Time: 30 minutes

Ingredients (for 4 persons):
- 1 big package of shredded squid (10 ounces)
- ⅓ -inch piece of ginger, sliced
- 2 cloves of garlic, sliced
- 4 stalks of scallion, chopped
- 2 tablespoons of red pepper powder
- 1 tablespoon of soy sauce
- 2 tablespoons of organic sugar
- 2 tablespoons of grapeseed oil
- 2 tablespoons of mirin
- sprinkle of pepper
- 1 tablespoon of Roasted sesame oil
- 1 teaspoon of roasted sesame seeds

Method:

1. Untangle the shredded squid and immerse it in water with ½ tsp of salt for about 2-3 minutes. Drain in a basket and set aside. This step helps remove any added MSG that might be present from the manufacturing process.

2. Line up all the seasonings near the pan: red pepper powder, sugar, soy sauce, mirin, and sesame oil. You want everything ready and within reach for quick cooking.

3. Heat a pan, add grapeseed oil, and quickly stir-fry the ginger until it becomes brown. Then, add the sliced garlic, chopped scallion, soy sauce, pepper, mirin, and, lastly, the red pepper powder. Do this step swiftly and turn off the heat.

4. Add the drained shredded squid to the pan, mix well, and turn the heat back on. Stir-fry for about 1-2 minutes, being careful not to overcook. Finally, add the sesame oil and sesame seeds.

Enjoy your clean-tasting shredded squid! Avoiding the use of red pepper paste and following these simple directions will ensure a delicious result.

Spicy Braised Chicken (Dakdoritang / 닭도리탕)

Total Time: 1 hour 30 minutes

Ingredients (for 4 persons):

- 2 pounds of chicken wings & thighs
- 2 medium potatoes, cut into chunks
- ⅔ big Spanish onion, cut into chunks
- 1 green pepper, chopped (1 ounce)
- ½ stalk of big green onion, chopped
- 4 cloves of garlic, minced
- 4 tablespoons of red pepper paste
- 2 tablespoons of red pepper flake
- 4 tablespoons of soy sauce
- 1 tablespoon of organic sugar
- ½ tablespoon of mirin
- 2 cups of water

Method:

1. Clean the chicken under cold running water and drain.

2. Make a marinade by mixing soy sauce, minced garlic, red pepper paste, red pepper flakes, and sugar.

3. Put the chicken parts and onion chunks into the marinade. Mix well with a big spoon.

4. Add water to the chicken and marinade mixture. Mix well, then transfer everything into a braising pot. Bring it to a boil over medium-high heat for 20 minutes, covered with a lid.

5. While the chicken is cooking, prepare the potatoes and green pepper as instructed above.

6. Add potato chunks and chopped green peppers to the pot and stir them in with a big spoon. Continue to braise for another 20 minutes, covered with the lid.

7. Now, start opening the lid periodically to allow some of the broth to evaporate and thicken the sauce. Cook for an additional 3-5 minutes. Occasionally, take some broth from the bottom of the pot and pour it over the top to ensure every ingredient absorbs the flavorful broth.

8. Serve hot, garnished with chopped big green onions.

Spicy Braised Tofu (Dubu Jorim / 두부조림)

Total Time: 30 minutes

Ingredients (for 2 persons):
- 1 box of tofu (10 ounces)
- Kosher salt: sprinkle
- 1 tablespoon of grapeseed oil
- ¾ cup of soy sauce
- 1 tablespoon of organic sugar
- 1 tablespoon of roasted sesame seeds, crushed
- 1 tablespoon of roasted sesame oil
- 1 tablespoon of red pepper powder
- 1 ½ stalks of scallion, sliced
- 3 cloves of garlic, diced
- ¼ cup of Water
- 1 ounce of radish slices or two sesame leaves: as cushion

Method:

1. Cut the tofu in half and slice it to a ⅓ -inch depth. Sprinkle some salt on the tofu slices. Set it aside.

2. After the salt has been absorbed into the tofu, use a paper towel to pat dry the tofu slices, removing excess water.

3. Heat a pan and add grapeseed oil. Stir-fry the tofu slices from step 2 until both sides turn a light-yellow color. Set it aside.

4. Prepare the marinade by mixing soy sauce, sugar, sesame seeds, sesame oil, red pepper powder, scallion, garlic, and water. Set it aside.

5. Place radish slices or sesame leaves as a cushion at the bottom of a saucepan. Add the tofu on top and pour the marinade over the tofu. Toss the marinade using a tablespoon and stir-fry for about 2 minutes over high heat. Reduce the heat to low, cover with a lid, and continue braising the tofu. From time to time, toss the marinade over the tofu pieces to enhance their flavor.

Tip:

This is a simple recipe but be cautious not to rely too much on the cushion (radish or sesame leaves). Burning the cushion can negatively impact the taste of the spicy braised tofu.

Steamed Egg (Gyeran Jjim / 계란찜)

Total Time: 20 minutes

Ingredients (for 2 persons):

- 2 eggs
- ½ teaspoon of Kosher salt
- 4 ounces of water (egg to water ratio 1:1)

Method:

1. In a bowl, break 2 eggs, add the salt, and pour in the water. Stir thoroughly using a whisk and then strain the mixture through a sieve. Set aside the container with the egg mixture.
2. In a 4-quart pot, bring water to a boil and place a steamer inside. Put the container with the egg mixture onto the steamer. Cover the pot with a glass lid.
3. Boil the water over medium heat for the first 8 minutes, then reduce the heat to low and steam for an additional 4-5 minutes. This will result in a soft and juicy steamed egg. The surface should change to a hardened texture with a pale-yellow color.

Tip:

When using a microwave oven, cover the container with plastic wrap and make several holes with a fork or chopstick. Microwave for about 2 minutes, then uncover and stir well to evenly mix the cooked and uncooked parts. Cover again with plastic wrap, make holes, and microwave for an additional 1 minute and 30 seconds to achieve a soft, pudding-like texture.

Stir-fried Anchovy (Myeolchi Bokkeum / 멸치볶음)

Total Time: 20 minutes

Ingredients (for 2 persons):
- 1 cup of small to medium-sized anchovies
- 1 tablespoon of grapeseed oil
- ⅓-inch piece of ginger, sliced
- 2 cloves of garlic, sliced
- ⅓ stalk of scallion, chopped
- 1 ½ tablespoons of organic sugar
- 1 tablespoon of mirin
- roasted and crushed sesame seeds
- roasted sesame oil

Method:

1. Begin by cleaning the anchovies. Sift away any broken pieces, but do not remove the heads and innards.

2. Heat a frying pan over strong heat, and swirl in the grapeseed oil. Once the oil is heated, add the sliced ginger. Stir-fry until the ginger turns brown.

3. Add the sliced garlic and chopped scallion to the pan. Continue stir-frying.

4. Quickly add the sugar, mirin, and anchovies to the pan. Stir-fry promptly.

5. Finish by sprinkling roasted and crushed sesame seeds and drizzling roasted sesame oil over the dish.

6. Serve your stir-fried anchovies.

Tip:

If you prefer a spicy version, use 1 tablespoon of red pepper, and increase the sugar by 0.5 tablespoons. Cook over medium heat.

If the anchovies you purchase are dried by heat, you can produce vitamin D by spreading them on a wide basket under direct sunlight for a day.

 Note

Anchovies are rich in nutrients like EPA, DHA, DMAE, purine, taurine, and vitamin B3 (niacin).

Stir-fried Dried Pollock (Bugeopo Bokkeum / 북어포볶음)

Total Time: 1 hour

Ingredients (for 2 persons):
- 1 small package of dried pollock (3 ½ ounces)
- 3 tablespoons of mirin
- 2 tablespoons of grapeseed oil
- 1 tablespoon of sesame oil
- 3 cloves of garlic, minced
- ⅓ -inch piece of ginger, sliced
- 2 tablespoons of soy sauce
- 2 tablespoons of organic sugar
- 2 tablespoons of red pepper powder
- 1 tablespoon of red pepper paste
- ½ tablespoon of roasted sesame seeds
- 1 green pepper chopped and sliced (1 ounce)

Method:

1. Trim the large and lengthy dried pollock by cutting it with scissors. Remove any small pieces that were created during the sieving process.

2. Mix the dried pollock with 3 tablespoons of mirin and let it soak.

3. Over medium heat, heat a frying pan with plenty of oil. Stir-fry the dried pollock until it turns a deep yellow color. Set it aside.

4. In the same pan, add sesame oil, garlic, ginger, soy sauce, sugar, red pepper powder, and red pepper paste. Mix them well and then add the stir-fried dried pollock from step 3. Continue stir-frying.

5. Toss the green pepper and sesame seeds on top of the stir-fried pollock mixture from step 4. Serve and enjoy.

Stir-fried Dry Squid Slice (Mareun Ojingeo Bokkeum / 마른오징어볶음)

**Total Time: overnight for soaking
and 20 minutes for cooking**

Ingredients (for 2 persons)

- 1 medium dry squid (3 ½ ounces)
- 1 tablespoon of grapeseed oil
- ½ teaspoon of Kosher salt
- 1 tablespoon of soy sauce
- 2 tablespoons of organic sugar
- 1 tablespoon of red pepper powder
- 1 ½ cloves of garlic, minced or cut into halves
- ¼ stalk of big green onion, chopped (2 ounces)
- 2 teaspoons of roasted sesame seeds
- 1 teaspoon of roasted sesame oil

Method:

1. Begin by soaking the dry squid in cold water overnight as a whole. This will allow it to soften and swell.

2. After soaking, thoroughly wash the softened dry squid, drain it, and then cut it into thin slices. Set it aside.

3. In a frying pan, heat the grapeseed oil. Once the oil is hot, add the garlic and stir-fry until it turns golden yellow. Then, add the chopped big green onion to infuse its flavor into the oil. Next, add the sliced squid and a pinch of salt, and continue stir-frying. Set this mixture aside.

4. Reduce the heat to low, and add the soy sauce, sugar, and red pepper powder to the pan. Once you see bubbles forming, add the prepared squid mixture from step 3. Cover the pan with a lid and let it simmer.

5. To serve, sprinkle roasted sesame seeds on top and drizzle with roasted sesame oil. This dish is a favorite from my childhood!

Stir-fried Glass Noodles & Vegetables (Japchae / 잡채)

Total Time: 1 hour 30 minutes

Ingredients (for 4~6 persons):

- 10 ounces of beef
- 10 ounces of glass noodles
- 10 large dried shiitake mushrooms
  (½ package)
- 10 sheets of dried wood ear mushrooms
- 1 bunch of spinach
- 1 large onion, sliced
- 1 large carrot, sliced
- 8 stalks of scallions
- 3 cloves of garlic
- ⅓ -inch piece of ginger
- 2 tablespoons of soy sauce
- sprinkle of Kosher salt
- 3 tablespoons of organic sugar
- 1 teaspoon of pepper
- ¼ cup of grapeseed oil
- 3 tablespoons of roasted sesame oil
- 1 tablespoon of roasted and crushed sesame
  seeds

Method:

1. Soak the glass noodles in lukewarm water for about an hour. Drain and set aside.
2. Rehydrate the shiitake mushrooms and wood ear mushrooms in cold water. Remove the stems from the shiitake mushrooms and slice the caps into halves. Separate and tear the larger wood ear mushrooms into smaller pieces. Set it aside.
3. Slice the beef and season it lightly with soy sauce, diced garlic, sugar, sesame seeds, and sesame oil. Stir-fry the beef until it's half-cooked, then add the shiitake mushrooms and wood ear mushrooms. Set it aside.
4. Blanch the spinach, rinse it under cold running water, drain, and squeeze out any

excess water. Quickly stir-fry the spinach and set it aside.

5. Cut the carrots, onions, and scallions into 2-inch lengthwise strips, then slice them a bit thicker. Stir-fry each vegetable separately while sprinkling some salt. Keep them separate on plates and let them cool. Do not mix them yet. Set it aside.

6. Boil water and cook the glass noodles until they become transparent and soft. Rinse them with cold running water and drain. Cut the noodles into 4–5-inch lengths. Set it aside.

7. In a heated frying pan, stir-fry a piece of ginger, then add the cooked noodles and continue to stir-fry, adding soy sauce, sugar, and sesame oil. Let it cool and set aside.

8. In a wide, large bowl, quickly mix the beef mixture (step 3), spinach (step 4), separately cooked vegetables (step 5), and the noodles (step 7) together with diced garlic, chopped scallions, sugar, sesame seeds, and half of the sesame oil. Add soy sauce to taste and add the remaining sesame oil last.

Tips:

• You can hold about 2.5 ounces of dry glass noodles between your thumb and middle finger. This can make measuring easier.

• If you mix the ingredients while any part is still warm, the colors may become dull, and the Japchae may spoil more quickly.

Stir-fried Glass Noodles with Bell Peppers (Pimang Japchae / 피망잡채)

Total Time: 1 hour

Ingredients (for 4~6 persons):
- 10 ounces of beef (flank steak)
- 10 ounces of glass noodles
- 10 large dried shiitake mushrooms (½ package)
- 2 large red bell peppers, sliced
- 2 large yellow bell peppers, sliced
- 2 large orange bell peppers, sliced
- 4 large green bell peppers, sliced
- 2 medium carrots, sliced
- 1 medium onion, sliced
- ½ stalk of big green onion, sliced
- 1 ½ cloves of garlic, sliced
- 1-inch piece of ginger
- 1 tablespoon of Kosher salt
- 2 tablespoons of soy sauce (adjust to taste)
- 3 tablespoons of organic sugar
- 1 teaspoon of pepper
- 3 tablespoons of roasted Sesame oil
- 1 tablespoon of sesame seeds
- ¼ cup of grapeseed oil

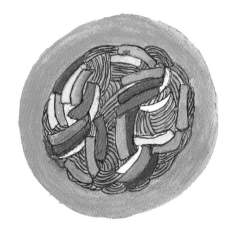

Method:
1. Begin by soaking the glass noodles in lukewarm water for 30 minutes to 1 hour, or until they become soft. Once softened, drain, and set aside.

2. Slice the beef flank into 2–3-inch lengths, cutting against the direction of the muscle fibers. Then, slice the beef into thin ¼-inch strips.

3. Heat a frying pan and add grapeseed oil. Stir-fry the ginger slices first, then add the beef strips along with minced garlic, salt, pepper, and mirin. Once cooked, set the beef aside.

4. Rehydrate the dry shiitake mushrooms until they become soft, remove the stems, and season with salt.

5. Slice the red, yellow, orange, and green bell

peppers into ¼-inch-thick strips. Quickly stir-fry each bell pepper separately with a pinch of salt, then cool each on separate plates. (Note: You mentioned green bell pepper but no spinach in the ingredients.)

6. Stir-fry the carrots, onions, and big green onions, then cool them on a wide plate. Set aside.

7. Cut the softened glass noodles into 4–5-inch lengths using scissors. Boil them in water until they become transparent, then rinse with cold water and drain using a basket.

8. Heat a frying pan and add grapeseed oil. Stir-fry the ginger, then add the boiled glass noodles. Keep stir-frying with soy sauce, sugar, and sesame oil, adding them gradually. Once done, cool the glass noodles.

9. In a large bowl, toss the cooled glass noodles with the beef (from step 2), shiitake mushrooms (from step 3), bell peppers (from step 4), and the carrot-onion mixture (from step 5). Mix quickly and add sesame seeds, sesame oil, and sugar. Season with soy sauce to taste and add a bit more sesame oil if needed.

Enjoy your Westernized Glass Noodles! It's a colorful and flavorful dish.

Stir-fried Glass Noodles with Mushrooms (Beoseot Japchae / 버섯잡채)

Total Time: 1 hour 30 minutes

Ingredients (for 4~6 persons):

- 10 ounces of beef flank, sliced thinly (⅛-inch thick)
- 10 ounces of glass noodles, cut to 5-inch lengths after soaking.
- ½ cup of large dried shiitake mushrooms (2 packages)
- ¼ cup of oyster mushrooms (2 packages)
- ¼ cup of maitake mushrooms (1 package)
- ¼ cup of beech mushrooms (1 package)
- 10 sheets of dried eye-ear mushrooms
- ½ bunch of spinach, blanched
- ½ bunch of chives, cut into 5-inch lengths
- 2 whole cucumbers, sliced
- 2 medium carrots, sliced
- 1 medium onion, sliced
- ¼ stalk of big green onion, sliced
- 3 cloves of garlic
- 2-inches pieces of ginger (larger amount than usual)
- 2 tablespoons of grapeseed oil
- 2 tablespoons of soy sauce
- 1 tablespoon of organic sugar
- 2 tablespoons of roasted sesame oil
- 2 teaspoons of crushed roasted sesame

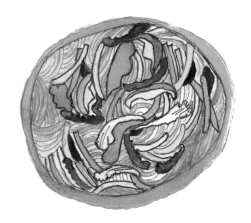

seeds
- ½ tablespoon of mirin

Method:

1. Slice the beef flank thinly and season it with 2 tablespoons of soy sauce, 1 tablespoon of sugar, 2 tablespoons of sesame oil, 2 teaspoons of sesame seeds, 1 tablespoon of chopped big green onion, ½ tablespoon of minced garlic, and ½ tablespoon of mirin. Set it aside.
2. Soak the dried Eye-Ear mushrooms in cold water to soften them. Also, soak the glass noodles in warm water for 30 minutes to 1 hour until they soften.

3. Handle each type of mushroom carefully. Stir-fry each type quickly to prevent sogginess:
- For raw shiitake mushrooms: Remove the stems and slice them thickly.
- For baby oyster mushrooms: Slice only the trunk part vertically.
- For maitake mushrooms: Cut out the bottom part and divide into several pieces.
- For beech mushrooms: Clean the bottom part.
- For swollen eye-ear mushrooms: Tear them into adequate sizes.

4. Stir-fry the seasoned beef from step 1. Once cooked, set it aside.

5. Stir-fry the cucumber, carrot, and onion with a pinch of salt separately. Set each aside.

6. In the same pan, stir-fry the ginger first for flavor, then add chives and spinach. Set each aside.

7. Boil the soaked glass noodles, rinse them with cold water, drain, and cut them into 5-inch lengths. In a heated frying pan, stir-fry ginger, add the glass noodles, and season with soy sauce, sugar, and sesame oil. Let it cool and set it aside.

8. Combine all the stir-fried ingredients (steps 3-7) in a large bowl. Add big green onion, garlic, sesame seeds, sesame oil, and sugar. Mix everything quickly and season with soy sauce. Add the sesame oil last.

This Stir-fried Glass Noodles with Mushrooms uses cucumber, which complements the soft texture of the mushrooms.

Stir-fried Golden-haired Squid (Ojingeo Silchae Bokkeum / 오징어실채볶음)

Total Time: 1 hour

Ingredients (for 4 persons):
- 1 package of dried squid thinly sliced (8 ounces)
- 3 tablespoons of grapeseed oil
- 2-inch pieces of ginger, thinly sliced
- 2 cloves of garlic thinly sliced
- 1~2 tablespoons of organic sugar
- ¼ teaspoon of mirin
- 1 tablespoon of sesame seed roasted & crushed
- 1 tablespoon of sesame oil

Method:

1. Open and spread-out dried squid that was packed compactly in a package. Maybe, the volume increases X3 more. Try to be patient when you open it. Do not hurry. Cut it in half with scissors.

2. Heat the frying pan. When heated, swirl around grapeseed oil, stir frying ginger to make ginger oil. Then, add garlic; the shape of these two will lose their shape by stir frying in a few seconds.

3. Turn down the flame as low as you can hardly see it, add dried squid and mix and stir well until oil smears into the squid all over. When the squid becomes transparent, add sugar and mirin and keep stirring. If the heat turns stronger, this recipe turns out to be a failure. Brown squid is not only ugly but also stiff, not tasty at all.

4. Sesame seeds and sesame oil last. They make golden-haired squid twinkle like stars.

Though very simple, you need some practice till you are used to making it. I hope this yellow-colored recipe becomes your favorite.

You don't need to add soy sauce or salt, because dried squid already has been seasoned during the manufacturing process.

Stir-fried Jiri Anchovies (Jiri Myeol Bokkeum / 지리멸볶음)

Total Time: 30 minutes

Ingredients (for 2 persons):
- ¼ cup of Jiri anchovy
- 1 tablespoon of minced onion
- 1 ½ cloves of garlic
- 1 tablespoon of organic sugar
- 3 tablespoons of mirin
- 2 tablespoons of grapeseed oil
- 1 tablespoon of roasted sesame oil
- sprinkle of roasted sesame seeds

Method:

1. Heat a pan, add half of the grapeseed oil, and stir-fry the jiri anchovies. Set them aside on a separate dish.

2. In the same pan, add the remaining half of the grapeseed oil and minced onion. Caramelize the onion by stirring. Continue to stir-fry while adding garlic, mirin, and half of the sugar. Finally, add the fried jiri anchovies from step 1.

3. Add the remaining sugar, roasted sesame oil, and a sprinkle of roasted sesame seeds.

4. Serve and enjoy!

Tip:

To fully savor the pure taste of jiri anchovies, avoid using soy sauce, malt extract, or green pepper in this dish. Cooking with minimum heat setting is essential for the best results.

Stir-fried Octopus (Nakji Bokkeum / 낙지볶음)

Total Time: 40 minutes

Ingredients (for 4 persons):
- 1 package of frozen octopus (1 ½ pounds)
- 4 tablespoons coarse sea salt (for rubbing octopus)
- 1 cup sliced onion
- ¼ stalk of big green onion, cut into 2-inch pieces
- 1 red or green pepper, sliced (1 ounce)
- ½ medium carrot, chopped
- ¼ cup of chopped cabbage
- ¼ potato, sliced
- 3 cloves of garlic, thinly sliced
- 1-inch piece of ginger, cut into two thick slices
- 2 tablespoons of grapeseed oil
- 2 tablespoons of soy sauce
- 2 tablespoons of red pepper powder
- 2 tablespoons of organic sugar
- 2 tablespoons of mirin
- ⅛ teaspoon of pepper
- 2 teaspoons of roasted and crushed sesame seeds
- 2 teaspoons of roasted sesame oil
- 1 teaspoon of pine nuts

Method:

1. Start by cleaning the octopus. Open the head and remove any internal contents, including the beak and ink sac. Rinse the octopus thoroughly.

2. Cut the octopus into 3–4-inch lengths, then rub them with 4 tablespoons of coarse sea salt. Rinse the salted octopus' pieces under cold running water and drain them using a colander. Set them aside.

3. Boil water in a pot and add the octopus along with a stick of ginger. Boil the octopus for no more than 10 seconds, then immediately remove it from the boiling water. Rinse the octopus with cold running water and drain again. Set it aside.

4. Heat a wok and evenly distribute the grapeseed oil. Once the oil is heated and starts to bubble, add ginger slices and stir-fry until they turn light brown.

5. Add the following ingredients in the specified order: sugar, soy sauce, garlic, sliced onion, the white part of the big green onion, chopped cabbage, sliced potato, red pepper, pepper, and red pepper powder. Quickly toss in the precooked octopus, the green part of the big green onion, and mirin. Mix and stir-fry everything swiftly. This step should be taken quickly for the best results.

6. Finally, add roasted sesame seeds and roasted sesame oil. Stir well to combine. Serve the dish and garnish it with pine nuts.

Stir-fried Oyster Mushrooms (Neutaribeoseot Bokkeum / 느타리버섯볶음)

Total Time: 20 minutes

Ingredients (for 4 persons):

- 8 ounces of oyster mushrooms
- 6 ounces of beef, shredded (like bulgogi)
- 1 tablespoon of grapeseed oil
- 2 tablespoons of oyster sauce
- 1 sweet onion
- ¼ stalk of big green onion
- ⅔-inch piece of ginger, sliced
- 2 cloves of garlic, sliced
- 1 tablespoon of cornstarch
- ¼ cup of water

Method:

1. Begin by warming a frying pan over strong heat. Swirl in the oil. Once the oil is heated, stir-fry the ginger, garlic, and shredded beef. Add the oyster sauce and mix everything well. Quickly toss in the onion and big green onion. Ensure the onion stays crispy.

2. Mix the cornstarch with water, then pour it over the contents from step 1. Stir well and serve the dish on a hot plate.

Stir-fried Pork belly (Samgyeopsal Bokkeum / 삼겹살볶음)

Total Time: 30 minutes

Ingredients (for 2 persons):

- 10 ounces of 3-layer pork belly
- ¼ cabbage (1 cup)
- 1 large, sweet onion
- ½ stalk of big green onion, chopped
- 1 tablespoon of red pepper paste
- 1 ½ tablespoons of red pepper powder
- 1 tablespoon of organic sugar
- 3 cloves of garlic, diced
- 1-inch piece of ginger, diced
- 2 ½ tablespoons of soy sauce
- 1 tablespoon of mirin
- ½ cup of water
- ½ teaspoon of roasted sesame oil
- paper towels (more than 10 sheets)

Method:

1. Heat a frying pan and stir-fry the pork belly on one side until it's slightly cooked. Push the pork belly to one side of the pan. Stir-fry the chopped big green onion to infuse its flavor into the pan, and then continue stir-frying the other side of the pork belly until all sides turn golden brown. Use plenty of paper towels to absorb the excess oil that comes out from the pork belly. This step is crucial to reduce excess oil and make the dish healthier.

2. Add sugar to the pan first, then pour in ½ cup of water. Gradually add red pepper paste, diced garlic, diced ginger, red pepper powder, soy sauce, and mirin. Mix all the ingredients well.

3. Add cabbage and sweet onion to the pan and continue stir-frying until they start to soften.

4. Drizzle roasted sesame oil over the dish just before finishing to enhance the flavor.

Stir-fried Rice Cake (Tteokbokki / 떡볶이)

Total Time: 30 minutes

Ingredients (for 4 persons):
- 8 ounces of rice cakes
- 3 rectangular pieces of fishcake (5 ounces)
- ¼ stalk of big green onion (2 ounces)
- 3 cloves of garlic
- 1-inch piece of ginger
- 10 large dried anchovies
  (approximately 1 ounce)
- 1 palm-sized piece of kelp
- ½ tablespoon of red pepper paste
- 1 ½ tablespoons of soy sauce
- 1 ½ tablespoons of organic sugar
- ½ tablespoon of corn syrup
- ½ tablespoon of red pepper powder
- sprinkle of pepper
- 5 cups of water

Method:

1. Begin by making the soup with large dried anchovies and kelp. Bring it to a boil and simmer for 10 minutes.

2. Cut each fish cake into two pieces, creating two right triangular shapes. Blanch them in the boiling soup, adding a piece of ginger for extra flavor.

3. Add all the remaining ingredients into the soup from step 1. Once the mixture starts boiling again, add the rice cakes and fish cakes. You can also add hard-boiled eggs and fried dumplings along with the rice cakes if desired.

Tips:

This simple recipe is a favorite among many! It's a crowd-pleaser.

If you don't have red pepper paste, there's another version of stir-fried rice cakes that uses various kinds of vegetables. It's known as "Royal Court Rice Cake."

Stir-fried, Yellow-dried Pollock (Hwangtaegui / 황태구이)

Total Time: 40 minutes

Ingredients (for 2 persons):
- 1 medium-sized yellow-dried pollock (2 ½ ounces)
- ¼ cup of water
- 1 tablespoon of roasted sesame oil
- 2 tablespoons of soup soy sauce
- 2 tablespoons of soy sauce
- 2 tablespoons of organic sugar
- ⅔ -inch piece of ginger, sliced (2 teaspoons)
- 1 clove of garlic, minced
- ¼ stalk of chopped big green onion (2 ounces)
- 2 tablespoons of mirin
- 1 teaspoon of roasted and crushed sesame seeds
- ½ teaspoon of pepper
- 2 tablespoons of grapeseed oil

Method:

1. Begin by cleaning the yellow-dried pollock. Remove the head, fins, and any small fishbones. Rinse the pollock under running water and drain. Place it on a plate, pour ¼ cup of water over it, and allow it to stand for 10 minutes. After confirming that it has softened, gently press it with your hands to remove any excess water from the flesh.

2. Score the skin on the backside of the yellow-dried pollock. This helps prevent it from shrinking during stir-frying.

3. Prepare the marinade by combining roasted sesame oil, soup soy sauce, soy sauce, organic sugar, sliced ginger, minced garlic, mirin, roasted and crushed sesame seeds, and pepper. Soak the pollock in this flavorful marinade and refrigerate it for at least 1 hour.

4. Heat a pan over high heat and add grapeseed oil. Once the oil is hot, add the marinated pollock, starting with the skin side down. When it's halfway cooked, reduce the heat to medium. This recipe features a distinctive blend of oil, mirin, ginger, and pepper, which gives it a flavor like bulgogi. Unlike some recipes, this one doesn't include red pepper or red pepper paste, resulting in a unique and delicious taste.

Sweet and Sour Pineapple Pork (Tangsuyuk / 탕수육)

Total Time: 8 hours for making watery starch and 1 hour for cooking

Ingredients (for 4 persons):
- 10 ounces of pork
- 2 teaspoons of soy sauce
- 2 teaspoons of mirin
- 1 teaspoon of salt
- 2 eggs
- 1 cup of watery starch (swollen with water)
- 2 cups of oil for frying

For the Sweet and Sour Sauce:
- ½ cup of water
- ⅓ cup of apple vinegar
- ⅓ cup of organic sugar
- 1 tablespoon of soy sauce
- 1 teaspoon of salt
- 6 tablespoons of watery starch (swollen with water)
- 4 ounces of pineapple
- half of a red bell pepper
- half of a green bell pepper

How to Make Watery Starch:

1. Combine corn or potato starch with water in a large bowl. Stir and mix thoroughly. Set it aside.

2. After several hours, the watery starch will settle down at the bottom of the bowl. Quietly pour out the top watery part and use only the bottom sediments. This is watery starch. Consider making 3-4 times more and storing it in your refrigerator for future Chinese cooking.

Method:

1. Cut the pineapple, red bell pepper, and green bell pepper into 1.5-inch squares. Set them aside.

2. Cut the pork into finger-sized pieces and season them with soy sauce, mirin, and salt. Set them aside.

3. After letting it sit for 15-20 minutes, whisk the eggs, 1 hour and add the seasoned pork to the watery starch. Gently mix them with your palm and fingers, ensuring that the pork is thickly coated with the mixture.

4. Heat the frying oil to 350 degrees Fahrenheit. Shallow fry the pork pieces (from step 3) twice. Set them aside.

5. Immediately, prepare the sweet and sour sauce. Boil water in a shallow frying pan and season it with soy sauce, salt, vinegar, and sugar. Add the pineapple and two squares each of red and green bell peppers. Stir and cook. Then, add half of the watery starch into the mixture. If the sauce is too thin, use the rest of the watery starch. A slightly thinner sauce looks more appealing and enhances the flavor.

 Notes

• These days, some people prefer enjoying "sweet and sour beef" rather than "sweet and sour pork." However, beef can often turn out tough. To tenderize beef, consider adding twice as much mirin and ½ teaspoon of ginger juice.

• At Chinese restaurants, you might hear youngsters asking, "Do you want 'Bu muk?' or 'JJik muk?'" It's a matter of personal preference whether you pour the sauce on top of your meat ("Bu muk") or dip your meat into the sauce ("JJik muk"). The author prefers "JJik muk" to savor the crispy taste of fried meat. Even a 1.5-year-old grandson enjoys holding and eating fried pork. If you choose "Bu muk," the sticky, gooey sauce might make it difficult for little ones to handle.

Sweet and Sour Wings

Total Time: 1 hour

Ingredients (for 6~8 persons):
- 1 ¼ pounds of party wings
  (¼ of 5 pound package)
- 2 tablespoons of coarse sea salt
- 1 cup of tomato ketchup
- 1 tablespoon of Worcestershire sauce
- ½ teaspoon of organic sugar
- 1 teaspoon of soy sauce
- 1 teaspoon of pepper
- ½ cup of crushed pineapple, drained.

Method:

1. Begin by sprinkling coarse sea salt or Kosher salt evenly over the chicken wings. 30 minutes later, thoroughly rinse the wings under cold running water to remove excess salt. Drain and set them aside.

2. In a large mixing bowl, combine the fully drained chicken wings with tomato ketchup, Worcestershire sauce, organic sugar, soy sauce, pepper, and crushed pineapple. Ensure that the marinade coats the wings evenly. Mix well to combine.

3. Preheat the oven to 400 degrees Fahrenheit.

4. Arrange the marinated chicken wings on a baking sheet or in a roasting pan. Roast the wings for about 30~40 minutes or until they are cooked through and reach your desired level of crispiness.

5. Remove the wings from the oven and let them cool for a few minutes. Serve these delicious party wings hot and enjoy!

These sweet & sour wings are sure to be a hit at any gathering or event. The combination of flavors, including the sweetness of pineapple and the tanginess of Worcestershire sauce, makes them a delightful treat.

Chapter II

Side Dishes

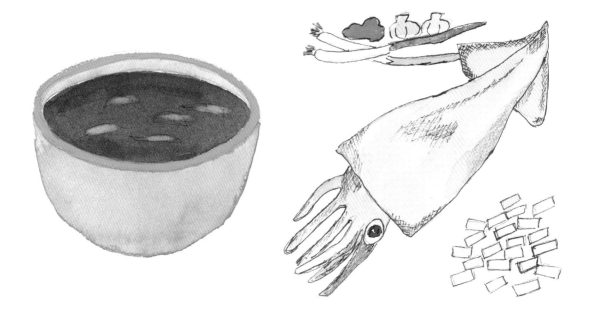

Acorn Jelly Making (Dotori Muk Ssugi / 도토리묵쑤기)

Total Time: 1 hour

Ingredients (for 4 persons):
- ½ cup of acorn powder
- 3 cups of water
- ¼ teaspoon of Kosher salt
- 1 drop of roasted sesame oil

Acorn Jelly Seasoning Sauce:
- 2 tablespoons of chopped big green onion
- 1 clove of garlic, diced
- 1 ½ tablespoons of soy sauce
- 1 tablespoon of soup soy sauce
- 1 tablespoon of red pepper powder
- ¾ tablespoon of roasted sesame oil
- 2 teaspoons of plum extract
- 1 teaspoon of malt syrup
- sprinkle of organic sugar
- sprinkle of sesame seeds

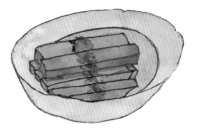

Method:

1. In a mixing bowl, thoroughly mix the acorn powder, salt, sesame oil, and 6 times the amount of water until there are no lumps of powder left.

2. Place a pot over medium heat and pour in the mixture from step 1. Bring it to a boil while continuously stirring to prevent lumps for more than 15 minutes. Initially, the liquid will look muddy.

3. When you start to see several little swirls on the surface, reduce the heat to low and continue stirring for about 15 more minutes. The mixture will become shiny and sticky, which is a sign that it's done.

4. Pour the cooked acorn mixture into a square glass or porcelain pan and let it cool at room temperature.

5. Once the acorn jelly is cooled, flip it over onto a cutting board and slice it using a jelly knife. Serve it with acorn jelly seasoning sauce.

Black Beans with Soy Sauce (Kongjorim / 콩조림)

Total Time: 1 hour

Ingredients (for 6 persons):
- 1 cup of black beans
- 3 cups of water
- 4 tablespoons of organic sugar
- 2 tablespoons of soy sauce
- 1 ½ tablespoons of soup soy sauce
- 3 tablespoons of bean-boiled water

Method:

1. Clean and wash the black beans thoroughly. Remove any possible little stones by rinsing the beans in two bowls of water alternately. Drain the beans and set them aside.

2. In a pot, add water and briefly boil the beans, covered with a lid. You'll know they're ready when there's no fishy smell when you taste them. Drain the beans using a strainer or basket but save some of the boiled water.

3. Over the black beans, pour the soy sauce, soup soy sauce, bean-boiled water, and half of the sugar.

4. Remove the lid and simmer until half of the liquid has evaporated.

5. Add the remaining sugar and continue to stir-fry and braise. The beans should have crumpled outer parts but remain soft on the inside. Store it in the refrigerator and enjoy this old-fashioned, nutritious side dish.

Braised Bellflower Roots (Doraji Namul / 도라지나물)

Total Time: 5 hours or overnight soaking and 30 minutes for cooking

Ingredients (for 4 persons):
- 2 cups of bellflower root (after soaking)
- 1 tablespoon of roasted sesame oil
- 2 tablespoons of Kosher salt (for rubbing) + 1 tablespoon (for seasoning)
- ⅛ stalk of big green onion, chopped
- 2 cloves of garlic, diced
- ⅛ teaspoon of pepper
- 4 tablespoons of water
- 1 teaspoon of roasted and crushed sesame seeds
- 1 teaspoon of red pepper strips (shilgochu / 실고추)

Method:

1. Soak the dried bellflower roots in cold water for over 5 hours (or overnight), then drain and set aside.

2. Rub the bellflower roots with 2 tablespoons of Kosher salt and soak them again in cold water for 1 to 2 hours to remove the bitter taste. **This step is important.** Drain and set aside.

3. Tear the thick roots into ⅛-inch slices, making them an edible size.

4. Heat a frying pan over strong heat, swirl in the sesame oil, add the bellflower roots, and stir-fry gently. When they become slightly transparent, add water and salt, cover with a lid, reduce the heat to medium-low, and braise for 10 to 12 minutes.

5. Sprinkle sesame seeds and red pepper strips, then serve.

Tips:

• Skipping or ignoring steps 1 and 2 may result in a bitter taste in the bellflower roots.

• For an even more delicious flavor, you can first stir-fry minced beef that has been marinated with bulgogi seasoning (with less sugar) before adding the bellflower roots.

Braised Bracken (Gosari Namul / 고사리나물)

**Total Time: 6 hours for soaking
and 1 hour for cooking**

Ingredients (for 4 persons):
- 2 palmful of dry bracken (2 ounces)
- 1 tablespoon of flour
- 2 ounces of diced beef (optional)
- 3 cloves of garlic
- ⅓-inch piece of ginger
- 1 stalk of scallion
- 1 tablespoon of soup soy sauce
- 1 tablespoon of mirin
- 1 teaspoon of grapeseed oil
- 1 tablespoon of crushed roasted sesame seeds
- 1 tablespoon of roasted sesame oil
- ⅓ cup of water

Method:

1. Immerse the dry bracken in cold water and let it swell for about 6 hours. Dissolve the flour in water along with the swollen bracken and wait for 20 minutes. This helps get rid of the unique fishy smell of bracken. 2 ounces of dry bracken will swell to 1 pound.

2. Boil the bracken in the flour-water mixture for 5 minutes. After boiling, wash it with cold water, drain, and let it cool. Immerse the cooled bracken again in cold water for 30 minutes. This step is essential to completely remove any residual smell.

3. In a pot, place the prepared bracken from step 2. Add garlic, ginger, soup soy sauce, mirin, sesame seeds, and grapeseed oil. Mix all the ingredients very well, preferably using your hands for thorough seasoning.

4. Initially, stir-fry the mixture over high heat. After 5 minutes, add scallion and sesame oil. Reduce the heat to low, cover with a lid, and braise for 20 minutes.

Serve and enjoy! My mother used to add diced beef at the beginning for added flavor, which is optional.

Braised Burdock Root (Ueong Jorim / 우엉조림)

Total Time: 40 minutes

Ingredients (for 4 persons):

- ½ pound of burdock root
- ½ teaspoon of Kosher salt
- 2 tablespoons of soy sauce
- ½ cup of water
- 2 tablespoons of organic sugar
- 1 tablespoon of mirin
- ⅛ -inch piece of ginger
- 3-4 cloves of garlic

Method:

1. Slice the burdock root as if you were making pencil cuts. Immerse the slices in 1% saltwater for 15 minutes.

2. Boil the sliced burdock root in water for 30 seconds, then wash them with cold running water and drain. Set them aside.

3. In a pot, mix soy sauce and water in a 1:2 ratio, then start boiling. Add the burdock roots over medium heat. When the water reduces by half, add sugar, mirin, ginger, and garlic. Reduce the heat to low. When there's almost no water left, briefly turn up the heat for 2 seconds until the surface becomes shiny. Turn off the heat and remove the ginger and garlic. Serve.

Tip:

Choose burdock root that is not stiff but moves softly when swung up and down.

Braised Lotus Roots (Yeongeun Jorim / 연근조림)

Total Time: 1 hour

Ingredients (for 4 persons):
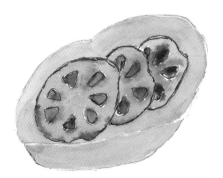
- ¼ pound of lotus root
- ½ teaspoon of Kosher salt
- 2 tablespoons of soy sauce
- 1 tablespoon of organic sugar
- ½ tablespoon of mirin
- ⅛ -inch piece of ginger
- 2 cloves of garlic

Method:

Preparation:

1. Start by selecting lotus root that doesn't have any scars on the surface and doesn't appear blackish in tint. Avoid purchasing pre-sliced and packaged lotus roots.

2. Peel off the skin of the lotus root using a potato peeler. Slice it into 0.25-inch-thick rounds.

3. Mix the Kosher salt into a ½ cup of water. Immerse the sliced lotus root in this saltwater solution for about 30 minutes. The saltwater will help prevent browning.

Cooking:

1. Bring a pot of water to a boil.

2. Blanch the lotus root slices in the boiling water for 3 minutes. This will give them a crispy and chewy texture.

3. In another pot, combine 2 tablespoons of soy sauce and 4 tablespoons of water. Bring this watery soy sauce mixture to a boil.

4. Add the blanched lotus root slices to the boiling watery soy sauce. Braise the lotus root over medium heat.

5. Once the watery soy sauce has reduced to about half, add the sugar, mirin, ginger, and garlic to the pot. Reduce the heat to low and continue braising until almost all the liquid has evaporated.

6. Just before turning off the heat, increase the heat to high and stir-fry the lotus root slices for a few seconds until they appear shiny and wet.

7. Remove from heat and serve your braised lotus root.

Chayote Salad (Chayote Muchim / 차요테무침)

Total Time: 30 minutes

Ingredients (for 4 persons):

- ¾ cup of chayote, thinly sliced in a fan shape
- 1 tablespoon of coarse sea salt
- 2 cloves of garlic, minced
- 1 stalk of scallion, chopped
- 1 tablespoon of red pepper powder (or red pepper paste)
- 1 tablespoon of rice vinegar
- 2 teaspoons of organic sugar
- 1 teaspoon of sesame seeds roasted and crushed
- 1 teaspoon of sesame oil roasted and crushed

Method:

1. Cut the chayote lengthwise into four pieces, then slice it into thin fan-shaped slices.

2. Salt the chayote slices and let them sit for 5 minutes. Then, rinse them with cold running water and drain. Squeeze out any excess moisture with your hands.

3. Season the chayote slices with scallion, garlic, red pepper powder (or red pepper paste), vinegar, and sugar.

4. Add sesame seeds and sesame oil to the mixture, stirring to combine.

Chicken Gravy

Total Time: 30 minutes

Ingredients (2 cups):

- ¼ cup of unsalted butter
- ½ cup of chopped onion
- 1 teaspoon of crumbled dry sage
- 1 teaspoon of crumbled dry thyme
- ¼ cup of all-purpose flour
- 2 ¼ cups of 2% milk
  (¼ cup is an additional amount)
- 2 cubes of chicken bouillon
- ½ teaspoon of Kosher salt
- ⅛ teaspoon of pepper
- ⅛ teaspoon of chopped parsley (for garnish)

Method:

1. In a medium-sized saucepan, melt the butter. Add the chopped onion, sage, and thyme. Sauté over medium-low heat for about 3 minutes. Stir in the all-purpose flour and whisk over low heat until a paste forms and the flour is lightly colored, which should take about 2-3 minutes.
2. Add the milk and bouillon cubes and whisk until smooth. Bring the mixture to a boil, then reduce the heat, and cook over low heat, stirring for about 10 minutes. Continue cooking until the mixture thickens, and the flour taste is cooked out, which should take an additional 5 minutes. If needed, you can slightly thin it with extra milk.
3. Strain the gravy to remove any lumps. Season it with salt and pepper. Garnish with chopped parsley.

Chives Salad (Buchu Muchim / 부추무침)

Total Time: 20 minutes

Ingredients (for 4 persons):
- ¼ bundles of chives (5 ounces)
- 1 tablespoon of red pepper powder
- 1 tablespoon of soy sauce
- 1 tablespoon of rice vinegar
- ½ tablespoon of organic sugar
- sprinkle of pepper
- 2 teaspoons of roasted & crushed sesame seeds
- 1 ½ tablespoons of roasted sesame oil

Method:

1. Cut 1-inch of the white part of the chives, trim the whole leaves, wash, and drain. Set it aside.
2. Cut the chives into 2-inch lengths and place them on a wide plate on the dining table.
3. Sprinkle the following ingredients over the chives in this order: red pepper powder, soy sauce, rice vinegar, sugar, a sprinkle of pepper, sesame seeds, and finally, a generous amount of roasted sesame oil.

 Note

Chives salad pairs exceptionally well with various beef recipes. It's best to prepare the marinade directly in front of the dining table just before serving.

Cold Cucumber Seaweed Soup (Oi Miyeok Naengguk / 오이미역냉국)

Total Time: 30 minutes

Ingredients (for 2 persons):

- ⅓ whole English cucumber, sliced (¼ cup)
- a handful of dry seaweed (¼ ounce)
- 1 tablespoon of diced onion
- 1 clove of garlic, minced
- 2 ½ cups of water
- ½ teaspoon of roasted sesame seeds
- ½ teaspoon of Kosher salt
- 1 tablespoon of soup soy sauce
- 2 tablespoons of organic sugar
- 3 tablespoon of apple vinegar

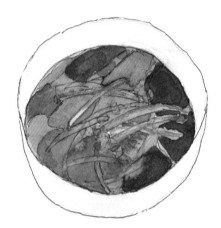

Method:

1. Let the dry seaweed swell in cold water for about 10-15 minutes. It will swell to an amount of 2 ounces.Wash it and drain. Blanch the seaweed in boiling water for 2 seconds, then wash and drain again. Cut the seaweed into 2-inch squares. Set it aside.

2. Slice the cucumber into ⅛ -inch-thick slices. In a bowl, mix the sliced cucumber with the seaweed from step 1, garlic, diced onion, salt, sugar, soup soy sauce, and vinegar. After letting it sit for 5 minutes, add cold water, mix well, and refrigerate.

3. When you're ready to serve, sprinkle roasted sesame seeds over the soup and toss in some ice cubes.

This refreshing summer soup is perfect for cooling down on a hot day. You can also add a drop of sesame oil just before eating if you like.

Egg Roll Slices (Hwangbaekjidan / 황백지단)

Total Time: 20 minutes

Ingredients (for 4 persons):
- 2 eggs
- ¼ teaspoon of Kosher salt
- ½ teaspoon of grapeseed oil

Method:

1. Break the eggs and separate the egg yolks and whites into two separate bowls. Add a pinch of salt to each bowl and gently beat them with a spoon. Be careful not to create too many bubbles, as you want a smooth surface.

2. Heat a frying pan over medium heat, add the grapeseed oil, and spread it evenly across the pan. Remove any excess oil.

3. Pour the egg white mixture into the pan, spreading it out as thinly as possible. Use a chopstick or utensil to lift a small part of the egg white to check if it's ready to flip. Once it's ready, flip it and cook for about 3 seconds on the other side. Then, transfer the cooked egg white to a wide plate.

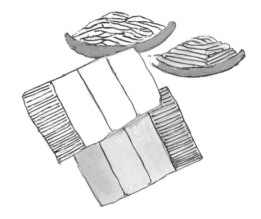

4. Repeat the same process with the egg yolk, spreading it thinly in the pan, flipping it, and cooking briefly.

5. After both the egg white and egg yolk are cooked, cut them into thin slices. These yellow and white egg slices will be used as garnish for tteoggug.

Fermented Dry Squid Slice with Rice (Bap Sikhae / 밥식해)

Total Time: 2 hours

Ingredients (for 8 persons):
- 1 cup of cooked rice
  (spread out on a basket and let it cool)
- 1 medium-sized Korean radish, thickly sliced
- 1 tablespoon of coarse sea salt
- 1 medium dry squid, thinly sliced and swollen in water (4 ounces)
- ¼ cup of red pepper powder
- 2 tablespoons of anchovy sauce
- 1 tablespoon of organic sugar
- 3 cloves of garlic, minced
- ⅛ -inch piece of minced ginger, minced
- ¼ stalk of big green onion diced and chopped

Method:

1. Brine radish slices with 3% salty water for one hour, then wash them with running water, drain, and squeeze thoroughly to remove excess water from the radish slices. Traditionally, people used to press a bag of radish slices under a flat rock to extract all the

2. Check if the dry squid slices have swollen enough, then squeeze them, drain, and set them aside.

3. Mix the cooked rice, brined radish slices from step 1, swollen dry squid slices from step 2, and red pepper powder together using your hands. Once these three ingredients turn red, add all the other ingredients.

4. Within 2 days, fermentation will begin. Store the container in the refrigerator. Over time, the sweet and sour flavors will develop, making it more delicious with each passing day.

Fermented Squid with Radish (Ojingeo Sikhae / 오징어식해)

Total Time: 2 hours

Ingredients (for 8 persons):
- ½ pound of squid (1 medium-sized squid)
- 1 medium-sized of Korean radish, thickly sliced
- 1 tablespoon of coarse sea salt (for brining)
- ¼ cup of red pepper powder (more amount than regular kimchi)
- 2 tablespoons of anchovy sauce
- 1 ½ cloves of garlic, diced
- ⅛ -inch piece of ginger, diced
- 2 tablespoons of organic sugar
- ½ teaspoon of Kosher salt
- malt syrup (use as needed)

Method:

1. Slice the radish into ¼-inch width, salt it with coarse sea salt for 3 hours. After salting is done, wash it in cold running water and drain. Put the radish into a thick cotton pouch and squeeze out excess liquid with your hands. Set it aside.

2. Prepare the squid by opening it lengthwise, removing the guts, and cleaning it thoroughly with clean water. Drain the squid and peel off the skin from the body part using a paper towel. Also, clean the sucking part of the 8 legs. Cut the body part into 3-inch-wide strips, turn them 90 degrees, and slice them to a ¼-inch thickness. Cut the legs into 3-inch lengths. Set them aside.

3. In a bowl, make the seasoning mixture by combining the red pepper powder, anchovy sauce, garlic, ginger, sugar, salt, and a bit of malt syrup (adjust to taste). First, mix the brined radish (step 1) with the seasoning mixture (step 3) until well combined. Then, add the squid (step 2) to the mixture and mix everything thoroughly.

4. Allow the kimchi to sit for 1-2 hours, and you'll notice that the redness becomes deeper. After several days, it will develop its unique hot, sweet, and sour taste. This dish is sometimes referred to as "<Rice Thief>" because it's so delicious that it makes you want to eat more rice.

Fried Gim with Rice Paper (Gim Bugak / 김부각)

Total Time: 30 minutes

Ingredients (for 5 persons):
- 5-6 sheets of gim, divided into 3 pieces each.
- 5 sheets of rice paper or 2 tablespoons of sweet rice powder
- Mixed with 7 ½ ounces of water (15 tablespoons)
- 2 cups of canola oil
- Roasted sesame seeds
- 2 tablespoons of organic sugar

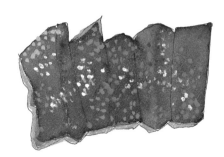

Method:

Using Rice Paper:

1. Begin by wetting the rice paper sheets and pressing them onto the surface of the gim (seaweed). Cut the gim into 3 pieces using scissors.

2. Heat the canola oil to 300 degrees Fahrenheit and quickly fry the prepared gim pieces. Fry until they become crispy.

3. Drain the excess frying oil from the fried gim and allow them to cool.

4. Once the gim has cooled down, sprinkle roasted sesame seeds and organic sugar evenly over the pieces.

Using Sweet Rice Powder:

1. Mix the sweet rice powder and water to make porridge. Dip one end of each gim sheet into the porridge and then fold the other end over it. Allow the gim to dry.

2. Heat the canola oil to 300 degrees Fahrenheit and fry the dried gim pieces until they become crispy.

3. Drain the excess frying oil from the fried gim and let them cool.

4. After the gim has cooled down, sprinkle roasted sesame seeds and organic sugar on top.

 Gim is a Korean term for seaweed, Laver, or nori. I prefer using the spelling 'gim' over 'kim' as it closely mirrors the original sound.

Fried Kelp (Dasima Twigak / 다시마튀각)

Total Time: 30 minutes

Ingredients (for 5 persons):
- 15 sheets of kelp
  (each sheet measuring 1 inch by 2 ½ inches)
- 2 cups of canola oil
- wet cloth towels
- 2 tablespoons of roasted sesame seeds
- 2 tablespoons of organic sugar

Method:

1. Begin by thoroughly rubbing and washing the surfaces of the kelp with wet towels. Repeat this process twice to ensure the kelp is clean. This step is crucial to remove any bitterness and excessive saltiness from the kelp. Set the cleaned kelp aside.

2. Heat the canola oil to 300 degrees Fahrenheit. Once the oil is hot, carefully fry the prepared kelp sheets until they turn crispy. This should be done quickly. Drain the excess frying oil from the fried kelp and allow them to cool.

3. After the fried kelp has cooled down, sprinkle roasted sesame seeds and organic sugar evenly over the sheets.

Enjoy your delicious fried kelp with sesame seeds and sugar!

Gim Mix (Gim Muchim / 김무침)

Total Time: 30 minutes

Ingredients (for 4 persons):
- 8 sheets of gim
- 1 tablespoon of soy sauce
- 1 tablespoon of soup soy sauce
- 1 ½ tablespoons of organic sugar
- 1 tablespoon of roasted sesame oil
- 1 tablespoon of roasted sesame seeds
- 3 cloves of diced green pepper, diced
- ½ red pepper, diced (1 tablespoon)
- 1 ½ cloves of garlic, minced

Method:

1. Roast the gim sheets using a portable grill. Roasting the gim eliminates any fishy smell and gives it a better texture.

2. Tear several sheets of roasted gim into small pieces, breaking them apart several times.

3. In a large bowl, mix the torn gim with soy sauce, soup soy sauce, and roasted sesame oil first. Then add roasted sesame seeds, diced green pepper, diced red pepper, and minced garlic. Stir well to combine. The watery ingredients will help prevent the gim pieces from clumping together.

Gim is a Korean term referring to seaweed, laver, or nori. I opt for the spelling 'gim' instead of 'kim' as it more accurately captures the original sound.

Later you will notice that I spell 'Gimbap' instead of 'Kimbap'. Like Bulgogi, Kimbap has gained recognition as a representative food name and has achieved worldwide acclaim.

Ginger Dressing

Total Time: 20 minutes

Ingredients (1 cup):

- ¼ cup of grapeseed oil
- ⅔-inch piece of chopped ginger, chopped
- ½ a medium-sized carrot, chopped
- 2 shallot or sweet onion, chopped
- ½ clove of garlic
- 1 tablespoon of local honey
- 1 teaspoon of roasted sesame oil
- 2 tablespoons of rice vinegar
- 2 teaspoons of soy sauce
- ¼ to ½ teaspoon of Kosher salt (adjust to taste)
- ⅛ teaspoon of white pepper

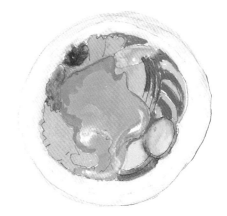

Method:

1. Roughly chop the ginger, carrot, and shallot or sweet onion and place them in a blender.
2. Add all the remaining ingredients to the blender.
3. Pulse and blend the ingredients 2 to 3 times or until you achieve the desired consistency, and everything is well combined.

4. Serve the ginger dressing with your mixed green salad. Enjoy!

Feel free to adjust the salt to your taste preference.

Greek Yogurt with Cucumber & Apple

Total Time: 20 minutes

Ingredients (for 4 persons):
- ⅓ cup of Greek yogurt (e.g., Chobani)
- 1 medium cucumber (7 ounces)
- 1 medium apple (¼ cup)
- 1 small lemon (2 ounces)

Method:

1. Cut the cucumber in half lengthwise, then cut one of the halves in half again, and finally cut into ½-inch-wide pieces. If you're using regular/slicing/English cucumbers, remove the seeded parts. If you're using Asian or Persian cucumbers, you can use the seeded parts. Place the cucumber pieces in a mixing bowl and set aside.

2. Cut the apple in half, then cut one of the halves in half again. Divide each piece into three and cut into ½-inch pieces. Mix the apple pieces with the cucumber pieces.

3. Cut the lemon in half and remove any seeds. Sprinkle the lemon juice onto the cucumber and apple pieces.

4. Pour the Greek yogurt onto the mixture from step 2 and serve.

This yogurt-based salad is a healthy side dish and can be enjoyed for breakfast as well. Chobani yogurt is known for its high protein content, making it a nutritious choice.

Green Garlic Stem Dish (Putmaneuldaemuchim / 풋마늘대무침)

Total Time: 30 minutes

Ingredients (for 4 persons):
- 2 stalks of green garlic stem, blanched (2 cups)
- ½ tablespoon of Kosher salt
- ½ tablespoon of crushed roasted sesame seeds
- ½ tablespoon of roasted sesame oil

Method:

1. Cut the green garlic stem into 1.5-inch lengths and blanch them in water (1 pint of water with ½ tablespoon of salt). After blanching, quickly wash them in cold water and drain. Set them aside.

2. In a bowl, place the drained green garlic stems from step 1. Add the salt and crushed sesame seeds and mix everything well. Finally, add the roasted sesame oil and give it a final mix.

This simple dish has a flavorful and clean taste. You can also enhance the taste by adding diced beef, diced braised beef, or diced red pepper.

Tips:

Here are some other ways to use green garlic:

A. **Green Garlic Stem with Ssamjang:** Cut the green garlic stem into 3-inch lengths and enjoy them with ssamjang. It will have a slightly spicy but flavorful taste.

B. **Mixture of Green Garlic White Roots:** Gather all the white roots of green garlic and mix them with soybean paste and plum extract in a 1:2 ratio.

C. **Green Garlic Leaves with Soybean Paste:** Mix 1 cup of soybean paste with 3 ounces of diced and minced green garlic leaves, 4 tablespoons of red pepper powder, and 6 tablespoons of sesame seeds.

D. **Green Garlic Leaves with Red Pepper Paste:** Mix 1 cup of red pepper paste with 3 ounces of green garlic leaves, 2 tablespoons of plum extract, and 1 tablespoon of soju. This can be used as a flavorful red pepper paste for dishes like Bibimbap.

E. **Green Garlic Kimchi:** You can make green garlic kimchi, but it's typically done once a year in May.

F. **Green Garlic Pickle (Preservation for 1 Year):** Combine soy sauce, vinegar, sugar, soju, and water in a 1:1:1:1:2 ratio. You can also add "soothe leek" when you start making the green garlic pickle.

Enjoy experimenting with these green garlic variations!

Jellyfish Salad (Haepari Naengchae / 해파리냉채)

Total Time: 1 hour

Ingredients (for 4 persons):
- ⅔ of 10 ounce-package of jellyfish (7 ounces)
- ½ of English cucumber, sliced
- ½ medium carrot, thinly sliced
- ½ red pepper, thinly sliced (½ ounce)
- 1 tablespoon of mustard sauce
- 2 tablespoons of water
- 1 teaspoon of Kosher salt
- 3 tablespoons of organic sugar
- 2 tablespoons of rice vinegar

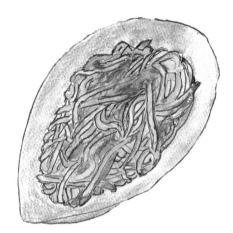

Method:

• Soak the jellyfish in warm water for 30 minutes, rubbing it to remove any excess sea salt. Drain and set aside.

• Blanch the jellyfish in boiling water, then rinse it in cold water and drain.

• Prepare a vinegary sweet water by mixing 1 cup of water, 3 tbsp sugar, and 2 tbsp vinegar.

• Squeeze out excess water from the jellyfish and immerse it in vinegary sweetwater. Set it aside.

• Slice the cucumber, carrot, and red pepper thinly.

• Make the jellyfish mustard sauce by adding 2 tbsp water to 1 tbsp mustard sauce.

• On a plate, arrange the jellyfish, cucumber, carrot, and red pepper. Drizzle the jellyfish mustard sauce on top. Serve.

Mustard Sauce Ingredients:
- 5 tablespoons of mustard powder
- 2 tablespoons of warm water
- 1 tablespoon of soy sauce
- ½ cup of organic sugar
- ½ cup of rice vinegar
- 1 tablespoon of Kosher salt
- roasted sesame oil (amount not specified,

adjust to taste)

Mustard Sauce Method:

1. In a mixing bowl, combine 5 tbsp of mustard powder with 2 tbsp of warm water. Stir well to form a smooth paste.
2. Add 1 tbsp of soy sauce to the mustard paste and mix thoroughly.
3. Gradually add ½ cup of organic sugar to the mixture, stirring until the sugar is fully dissolved.
4. Pour in ½ cup of rice vinegar and continue to mix until all the ingredients are well incorporated.
5. Season the sauce with 1 tbsp of Kosher salt, adjusting the saltiness to your preference.
6. Optionally, add roasted sesame oil to the sauce, adjusting the amount to your liking. This adds a nice nutty flavor to the sauce.
7. Once the sauce is well-mixed and seasoned to your taste, it's ready to use.

You can store any extra mustard sauce in the refrigerator and use it for various dishes. Enjoy your mustard sauce with the Jellyfish Salad!

Korean Style Dipping Sauce (Ssamjang / 쌈장)

Total Time: 30 minutes

Ingredients (for 4 persons):
- ¼ cup of soybean paste
- 2 tablespoons of red pepper powder
- 1 to 2 tablespoons of organic sugar or honey
- ¼ onion, diced
- 1 stalk of scallion, diced
- ½ green pepper, diced
- 2 teaspoons of roasted crushed sesame seeds
- 1 tablespoon of sesame oil

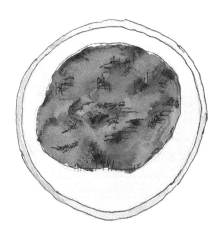

Method:

1. Homemade soybean paste can vary in saltiness from family to family, so the amount of sugar needed depends on the saltiness of your soybean paste. If you're using store-bought soybean paste, adjust the sugar accordingly. Add the sugar to the soybean paste.
2. Dice the onion, scallion, and green pepper.
3. In a small bowl, combine the soybean paste, diced onion, diced scallion, red pepper powder, and diced green pepper.
4. Once these ingredients are evenly mixed, sprinkle in the sesame seeds and drizzle plenty

of sesame oil over the mixture.

 Note

Ssamjang is a versatile condiment that pairs well with BBQ meat and various vegetables. There are different ways to adjust the ratio of soybean paste and red pepper paste based on personal preference, and you can even experiment with using honey instead of sugar or different types of soybeans paste. Get creative and make your own unique ssamjang!

Kyung's Apple Salad

Total Time: 20 minutes

Ingredients (for 4 persons):
- 1 medium apple
- 1/10 cabbage, sliced
- ½ cucumber, sliced
- 2 eggs, boiled and chopped
- 2 tablespoons of walnut, chopped and diced
- 1 tablespoon of mayonnaise
- 1 tablespoon of whipped cream
- 1 tablespoon of white vinegar (for boiling egg)

Method:

1. Cut the apple into quarters, remove the seeded part, cut each quarter into two lengthwise pieces. Partially peel the skin and then cut them into 3 or 4 pieces.
2. Slice the cabbage to a thickness of ¼-inch, wash it in cold water, and drain. Set it aside.
3. Slice the cucumber thickly. Set it aside.
4. Boil the eggs in boiling water (add 1 tsp vinegar) for 8 minutes. After boiling, wash them in cold water and peel off the shells. Rinse them again. Set it aside.

5. Chop and dice the walnuts. Set them aside.
6. In a bowl, mix the apple, cabbage, cucumber, boiled and chopped eggs, and chopped walnuts. Add the mayonnaise and whipped cream. Mix everything well. Cover the bowl with plastic wrap and refrigerate for 30 minutes to 1 hour.

This recipe is a creation of my own. It is delicious.

Pan-fried Eggplant (Gaji Gui / 가지구이)

Total Time: 20 minutes

Ingredients (for 4 persons):
- 2 medium-sized eggplants
- 1 ½ tablespoons of soy sauce
- 1 tablespoon of organic sugar
- 1 tablespoon of rice vinegar
- 2 cloves of garlic, minced
- 2 tablespoons of unsalted butter

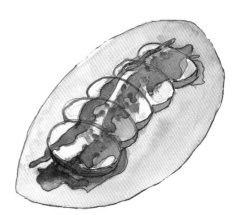

Method:

1. Slice whole cleansed eggplants 1-inch thick, steam halfway on bamboo steamer stack. After the eggplant smell comes out, turn the heat off, rinse the eggplants with cold water. Drain.

2. Make marinade with soy sauce, sugar, vinegar, and garlic.

3. Over strong heat, heat the pan. Once pan is heated, reduce the heat to weak, pan frying steamed eggplant with butter till both surfaces of # 1 eggplants turn yellow.

4. Coat both sides of eggplants with marinade and keep stir-frying each side. Serve.

Parboiled Spinach Dish (Sigeumchi Namul / 시금치나물)

Total Time: 20 minutes

Ingredients (for 4 persons):

- 1 bundle of spinach
  (separate each stem from the root) (1 pound)
- 1 tablespoon of Kosher salt
  (for parboiling)
- 1 tablespoon of soup soy sauce
- 2 cloves of garlic, minced
- 2 teaspoons of roasted and crushed sesame
  seeds
- 2 teaspoons of roasted sesame oil

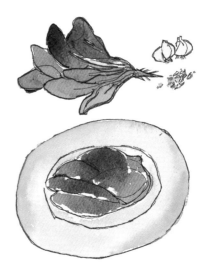

Method:

1. Clean and wash the spinach, paying special attention to the root part. Drain and set aside.
2. In a pot, bring 2 quarts of water to a boil and add salt. Parboil the spinach, flip it over, then quickly place it under cold running water to wash and drain. Set it aside.
3. Gently squeeze the spinach to remove excess water, then place it in a medium-sized bowl. Add soup soy sauce, minced garlic, sesame seeds, and roasted sesame oil. Mix well and serve.

 Note

Spinach is a food with high oxalate/oxalic acid content. Boiling vegetables can reduce the soluble oxalate content from 30~90%. Steaming can also reduce oxalate content, but boiling is more effective. (Source: Journal of Agricultural and Food Chemistry, US National Library of Medicine.)

This version includes information about oxalates for those interested in the nutritional aspect of the recipe.

Radish Salad with Sugar & Vinegar (Mu Saengchae I / 무생채 I)

Total Time: 20 minutes

Ingredients (for 4 persons):
- 1 small Korean radish, peeled and thinly sliced
- ½ tablespoon of sea salt or Kosher salt
- 2 cloves of garlic, minced
- ¼ teaspoon of ginger juice
- 1 ½ stalk of scallion
- 1 tablespoon of rice vinegar
- 2 teaspoons of red pepper powder
- 1 tablespoon of organic sugar
- 2 teaspoons of roasted and crushed sesame seeds

Method:

1. Start by peeling the radish and slicing it into thin strips, approximately ⅛ -inch thick. Avoid making the strips too thin, as they might clump together and form balls, making it challenging to marinate. After slicing, immerse the radish strips in cold water for about 30 seconds. Drain them and set aside.

2. Place the radish strips in a large bowl, add salt, and mix them by hand. Allow them to sit for 5 minutes to draw out excess moisture. This step is known as salting.

3. After 5 minutes, squeeze out the excess water from the radish strips.

4. Add minced garlic, ginger juice, scallion, rice vinegar, red pepper powder, and sugar to the radish strips. Mix everything together thoroughly. Finally, add the roasted and crushed sesame seeds and give it a final mix.

 Notes

• There are two more methods to make radish salad. One involves adding sugar without vinegar, where salt is omitted, and instead, shrimp sauce or anchovy extract is used. The other method omits sugar and vinegar entirely when making radish salad. I'll post the details of the second method on this site later.

• There are two primary ways to prepare raw vegetable side dishes (uncooked vegetable side dishes):

• Dressed with vinegar: Examples include radish, cucumber, onion, shallot, wild garlic, lettuce, chayote, arugula, and those used in salads. Seasoning is the same as in the radish salad.

• Quick salting: This method is used for cabbage (quick fresh kimchi), spring greens (quick fresh side dishes), and chives. Seasoning involves salt, soybean paste, red hot paste, or soup soy sauce.

Radish Salad without Sugar & Vinegar (Mu Saengchae II / 무생채 II)

Total Time: 20 minutes

Ingredients (for 4 persons):
- 1 small Korean radish (2cup)
- ½ tablespoon of coarse sea salt or Kosher salt
- 1 tablespoon of red pepper powder
- 2 cloves of garlic, minced
- 1 ½ stalks of scallion, sliced
- 1 tablespoon of roasted sesame seeds, crushed
- ½ tablespoon of roasted sesame oil

Method:

1. Cut the radish into 3-inch lengthwise sections, then slice each section into ⅛ -inch width slices.
2. Sprinkle salt evenly over the radish slices. Do not rinse off the salt; it's used for seasoning in this recipe.
3. Slice the scallion or big green onion to match the length and width of the radish slices.
4. Mix the salted radish slices from step 2 with minced garlic and red pepper powder. Add sliced scallion or big green onion, roasted sesame seeds, and roasted sesame oil. Mix everything thoroughly.

Enjoy this radish salad as a delicious accompaniment to rice. It offers a delightful combination of crispness and juiciness. Whether to add sugar and vinegar is a personal preference; some enjoy it with a touch of sweetness and acidity, while others prefer it without.

Raw Crab Marinated in Soy Sauce (Ganjang Gejang / 간장게장)

Total Time: 3 hours

Ingredients (for 6 persons):
- 6~7 raw crabs, live and healthy (2 pounds)
- 1 ½ quarts of water (equivalent to 6 cups)
- 2 ½ cups of soy sauce
- 1-inch piece of ginger, chopped
- 6 cloves of garlic, whole
- 2 green peppers (2 ounces)
- 2 medium onion
- 1 apple thinly sliced
- 5 palm-sized pieces of kelp
- 1 dried whole red pepper
- ¼ cup of rice syrup

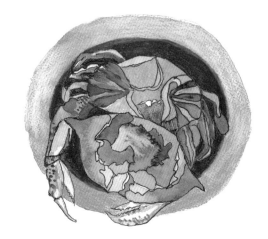

Method:

I used to struggle handling live crabs. But, these days, I picked up a special tip from Mangchi: you can induce sleep in live crabs by freezing them for 2 hours.

While the crabs are asleep, you can complete the entire procedure. However, be cautious; if you freeze them for over 2 hours or overnight, even if you thaw them properly, mucus can come out from their flesh, and your "ganjang gejang" might turn out to be a failure. If you freeze them for less than 2 hours, for example, just 1 hour

in a hurry, the crabs may wake up and become active in your sink.

1. In a deep pot, pour water and add soy sauce, ginger, garlic, green pepper, onion, apple, kelp, dried red pepper, and rice syrup. Cover the pot with a lid and boil it for 20 minutes over medium heat.

2. Remove the kelp and continue boiling on low heat (simmer) for about 1 hour.

3. After the mixture has somewhat cooled, filter

out the liquid through a cotton cloth placed on a basket. Allow it to cool completely.

4. Take out the sleeping crabs, place each one lying down on a cutting board, cover-part on top. Separate the top from the body-part by pulling. Remove the gill, sandbag, intestine, antennae, and eyes. Don't forget to remove any sand from its mouth by rubbing it with two fingers.

5. After the cleaning procedure is complete, place the body part back into its original position on the crab.

6. In a large-mouthed bottle, arrange the crabs from step 5 with the cover-part facing down, layer by layer. Pour the liquid from step 3 over the crabs. A heavy round stone can be helpful to keep all the white parts of the crabs submerged. No crab part should be above the surface. After 1 day, it's ready to serve.

Scallion Salad (Pamuchim / 파무침)

Total Time: 20 minutes

Ingredients (for 4 persons):
- 1 bundle of scallions (1 cup)
  (6~8 stalks are in one big bundle of scallion.)
- 1 tablespoon of rice vinegar
- 1 teaspoon of soy sauce
- ½ teaspoon of organic sugar
- 2 teaspoons of red pepper powder
- 1 teaspoon of roasted and crushed sesame seeds
- sprinkle of Kosher salt
- sprinkle of pepper
- 1 ½ tablespoons of roasted sesame oil

Method:

1. Slice the scallions into pieces that are 4-5 inches long, then cut each piece in half and slice them very thinly. If needed, you can find a special knife for cutting scallions at a Korean market.

2. Wash the sliced scallions in cold water, leaving them in the water for about 30 seconds, then drain and set them aside.

3. After the scallion slices are fully drained, spread them out on a large plate in a relaxed manner. Do not press them down.

4. Just before serving, sprinkle the rice vinegar, soy sauce, salt, sugar, pepper, red pepper powder, and sesame seeds over the scallions in that order. Finally, drizzle the sesame oil on top.

This scallion salad is a popular side dish, perfect for accompanying BBQ. It enhances the flavor of the meat and aids digestion. If available, you can use big green onions, which also provide great flavor.

Seasoned Crab (Yangnyeom Gejang / 양념게장)

Total Time: 1 hour

Ingredients (for 4 persons):

- 4 frozen whole crabs (2 pounds)
- ¼ cup of soy sauce
- 2 red-hot pepper (2 ounces))
- 2 tablespoons of organic sugar
- 1 tablespoon of corn syrup
- ½ teaspoon of salt
- ½ tablespoon of mirin
- 3 cloves of garlic, minced
- 2-inch piece of ginger, minced
- ⅛ stalk of big green onion chopped and diced
- 1 green pepper chopped and diced. (1 ounce)
- sprinkle of pepper
- 1 teaspoon of sesame seeds

Method:

1. Clean the top shell cover and body part of the frozen crabs with a stiff brush, making sure not to miss any hidden parts.

2. Trim the tips of the legs and remove sharp edges from the shell cover using scissors.

3. Separate the cover and body part, and clean out the gill, sandbag, intestine, antennae, and eyes.

4. Rinse thoroughly under slowly running water after the cleaning procedure is complete. Ensure that you remove any sand from their mouths by rubbing with your fingers.

5. Divide the body part into 2 and then into 4, creating a total of 8 pieces per crab. Cover each claw with cloth and break the hard shell halfway with the back of a knife. This allows the marinade to penetrate the flesh easily.

6. In a large bowl, prepare the marinade by combining soy sauce, red pepper powder, sugar, corn syrup, salt, mirin, minced garlic and ginger,

big green onion, green pepper, red pepper, pepper, and sesame seeds.

7. Keep the marinated crabs in the refrigerator and serve them after 1 hour. Enjoy your "Babdoduknom" or "Rice Thief"!

 Notes

• The seasoned crab will have its optimum taste after one day.

• Consider using disposable gloves and a vinyl apron to avoid staining your clothes with crab juices.

• In the freezer section of a Korean market, you can find pre-cleaned crabs for seasoned crab, which can be more convenient. If using these, you can skip steps 1 to 4 and begin from step 5.

Seasoned Cucumber and Bellflower Roots
(Oi Doraji Muchim / 오이도라지무침)

Total Time: 1 hour

Ingredients (for 4 persons):
- 4 medium-sized cucumbers
  (2 cups / 1 pound)
- ½ cup of raw bellflower root
- 2 tablespoons of Kosher salt
- ½ cup of sliced sweet onion
- ¼ cup of red pepper paste
- ¼ cup of red pepper powder
- 3 cloves of garlic, minced
- 5 tablespoons of rice vinegar
  *1 tablespoon of rice vinegar for reducing the
    bitter taste of bellflower root
  *4 tablespoons of rice vinegar for seasoning
- 6 tablespoons of organic sugar
  organic sugar 2 tablespoons are for draining
  water, and 4 tablespoons are for seasoning
- 1 tablespoon of roasted sesame seeds

Method:

1. Begin by rubbing the surface of the cucumbers with salt, then cut off both ends, and rinse them with cold water. Set them aside.

2. Soak the raw bellflower roots in water with 1 tablespoon of vinegar, 1 tablespoon of salt, and 1 tablespoon of sugar. Allow them to stand for about 40 minutes. The vinegar is used to remove the bitter taste of the bellflower roots, while the salt and sugar help draw out excess water from the vegetables. Set it aside.

3. At the same time as step 2, cut the cucumbers in half, slice them to a thickness of ⅛-inch, and soak them in water with 1 tablespoon of salt and 1 tablespoon of sugar for 40 minutes. Set them aside.

4. After 40 minutes, wash the bellflower roots and cucumbers together under cold running water, drain them, and squeeze out excess water using your hands.

5. In a mixing bowl, combine the squeezed bellflower roots and cucumber slices with the sliced onion, red pepper paste, red pepper powder, 4 tablespoons of sugar, 4 tablespoons of vinegar, and minced garlic. Mix everything thoroughly.

6. Finish by garnishing the dish with roasted sesame seeds. Serve and enjoy.

Seasoned Dried Radish (Mu Mallengi Muchim / 무말랭이무침)

**Total Time: 2 days for soaking
and 30 minutes for seasoning**

Ingredients (for 4 persons):

- ¼ cup of dried radish
- 2 cups of soy sauce
  (for soaking the dried radish)
- 1 ½ tablespoons of organic sugar
- 1 tablespoon of malt extract
- 1 tablespoon of mirin
- 3 tablespoons of red pepper powder
- 3 cloves of garlic, minced
- 1 tablespoon of roasted sesame seeds
- scallion: To be added just before serving
- roasted sesame oil:
  To be added just before serving

Method:

1. Place the dried radish in a 2-pint bottle, packing it tightly so that it is fully immersed in the soy sauce. Place something heavy on top to ensure the dried radish remains submerged in the soy sauce.

2. After two days, drain the dried radishes. Squeeze them out, little by little, while wrapped in a burlap cloth until they are almost dried, and there is no soy sauce left in them.

3. In a large bowl, combine the dried radishes from step 2 with the organic sugar, malt extract, mirin, red pepper powder, minced garlic, and roasted sesame seeds. Mix by rubbing the ingredients with your hands repeatedly to allow for deep seasoning.

4. Just before serving, add a small amount of sliced scallion and roasted sesame oil for extra flavor.

Seasoned Mung Bean Sprouts (Sukju Namul Muchim / 숙주나물무침)

Total Time: 20 minutes

Ingredients (for 2 persons):
- 1 package of mung bean sprouts (thoroughly washed three times) (1 ½ cups)

 Note

Mung bean sprouts weigh 3.7 ounces per cup volume.
Therefore, 1 ½ cups of mung bean sprouts yield 5.6 ounces.

- ⅔ stalk of scallion, chopped
- 2 cloves of garlic, minced
- 1 teaspoon of soup soy sauce
- 1 teaspoon of roasted and crushed sesame seeds
- 1 teaspoon of roasted sesame oil

Method:

1. Begin by boiling the mung bean sprouts for approximately 5 minutes. Once boiled, drain them, and set them aside to cool.
2. When the mung bean sprouts have partially cooled down, add minced garlic, chopped scallion, soup soy sauce, and sesame seeds. Mix these ingredients thoroughly using your bare hands. Add the roasted sesame oil as the final touch.

Additional Notes:

• This recipe is incredibly simple and easy to follow. You can adjust the spiciness by adding a desired amount of red pepper powder according to your taste.

• When preparing vegetable side dishes, there are primarily three methods of cooking, and the seasoning is the same for each:

  ○ Boiling (4-5 minutes): Suitable for mung bean sprouts, soybean sprouts, cold greens, and sow thistle.

  ○ Blanching: Suitable for spinach, watercress, water dropwort, crown daisy, and collard greens.

  ○ Steaming: Suitable for eggplant, asparagus, and cabbage leaves.

Seasoned Soybean Paste (Gang Doenjang / 강된장)

Total Time: 30 minutes

Ingredients (for 3 persons):

- 2 tablespoons of soybean paste
- ¼ tablespoon of red pepper paste
- 1 tablespoon of mirin
- ¼ of sweet onion, chopped
- ¼ of zucchini, diced and chopped
- ¼ cup of shiitake mushrooms, diced
- ¼ box of tofu, cut into 1-inch cubes or mashed
- 1 tablespoon of big green onion, diced
- 1 green pepper, diced. (1 ounce)
- 1 ½ cloves of garlic, minced
- ¼ cup of rice water

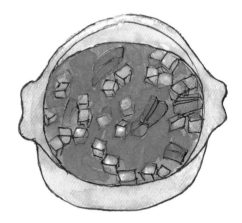

Method:

1. When steaming zucchini, place a metal or porcelain bowl on the bamboo shelf side by side, near the zucchini leaves in a bamboo steamer.
2. After the water in the steaming pot boils, steam the ingredients mentioned in step 1 for 15 minutes. This will result in the right amount of water inside the seasoned soybean paste, giving it a smooth texture without being overly salty.

 Notes

- You can accompany seasoned soybean paste with raw sesame leaves, swollen seaweed and lettuce. It pairs well with bokkeum red pepper paste.
- Gangdoenjang has less water compared to jjigae, despite using the same amount of soybean paste.

Seasoned Soybean Sprouts (Kongnamul Muchim / 콩나물무침)

Total Time: 20 minutes

Ingredients (for 3 persons):

- 1 package of soybean sprouts, cleaned, washed, and drained. (12 ounce)
- 2 cups of water
- 1 stalk of chopped scallion
- ½ clove of garlic, diced
- ¼ tablespoon of Kosher salt
- 2 teaspoons of red pepper powder
- 1 teaspoon of roasted sesame seeds
- 1 tablespoon of roasted sesame oil

Method:

1. Heat a 4-quart saucepan, place the soybean sprouts in the pot, add 2 cups of water, and cover with a lid. Boil for about 3-5 minutes, or until you can smell the soybean sprouts. Turn off the heat and wait for 30 seconds. Rinse the boiled soybean sprouts with cold running water and drain them.

2. In a deep bowl, put the drained soybean sprouts. Add salt, garlic, sesame oil, and red pepper powder. Mix all the ingredients using chopsticks. Finally, add the chopped scallion and sesame seeds, and give it one more gentle mix with the chopsticks. The result should be very crispy and delicious!

Steamed Eggplant (Gaji Namul / 가지나물)

Total Time: 30 minutes

Ingredients (for 4 persons):
- 3 long eggplants (10 ounces)
- 1 stalk of scallion, chopped
- 2 cloves of garlic, diced
- 1 tablespoon of soup soy sauce
- ½ teaspoon of sesame seed roasted crushed
- ½ tablespoon of sesame oil roasted crushed
- ½ teaspoon of red pepper (optional)

Method:

1. Choose straight shaped ones and with no scars on the skin. Cut the top out and wash, and drain. Set it aside.

2. Cut in half and cut again by four. Steam them in a bamboo steamer by 2 or 3 layers. Let the purple color face the sky for 5 minutes. Do not steam the eggplants too long so as not to lose their pretty purple colors. Wash them under cold running water by keeping eggplant in the steaming rack. This way you can prevent breaking the soft flesh of eggplant. Drain.

3. After fully draining, season with scallion, garlic, soup soy sauce, sesame seeds, and sesame oil. When you mix seasonings, treat the eggplant gently, otherwise the soft tissue of eggplant will be broken and look ugly. Serve as a side dish.

Usually, we blanch vegetables when we cook <namul>. But we steam eggplant, asparagus, and cabbage leaves.

Stir-fried Chayote (Chayote Bokkeum / 차요테복음)

Total Time: 20 minutes

Ingredients (for 4 persons):
- 1 large chayote, sliced (¾ cup)
- 2 cloves of garlic, minced
- 1 ½ stalk of scallion, chopped
- ½ red pepper, sliced
- 1 teaspoon of Kosher salt
- 1 tablespoon of grapeseed oil
- 1 teaspoon of roasted sesame seeds
- 1 teaspoon of roasted sesame oil

Method:

1. Cut the chayote into four pieces lengthwise, remove the seed part, but keep the skin on. Slice each piece to about ⅛ -inch thickness. Be careful not to make the slices too thin.
2. Heat a frying pan until hot, then add the oil. Once the oil is heated, add the sliced chayote, garlic, and salt. Stir-fry this mixture until the chayote slices become half-transparent. Add the scallion and red pepper, then finish with sesame seeds and sesame oil. Serve.

This dish retains its flavor even after cooling and develops a unique taste with its chewy texture.

Stir-fried Eggplant (Gaji Bokkeum / 가지볶음)

Total Time: 20 minutes

Ingredients (for 2 persons):

- 2 medium eggplants (2 cups)
- 1 tablespoon of Kosher salt
- 1 onion, sliced
- 1 tablespoon of grapeseed oil
- 1 ½ cloves of garlic, minced
- ½ stalk of scallion, chopped
- 1 tablespoon of soy sauce
- ½ tablespoon of organic sugar
- 1 teaspoon of roasted sesame seeds, crushed
- ½ teaspoon of roasted sesame oil

Method:

1. Begin by cutting the eggplants in half lengthwise and then slicing them into ½-inch-thick pieces sideways.

2. In a bowl, brine the sliced eggplants with Kosher salt. After about 10 minutes, rinse the eggplant slices under running water and gently squeeze them with a paper towel to remove excess moisture. Set them aside.

3. Heat a thick frying pan and swirl around the grapeseed oil. Once the oil is hot, add the prepared eggplant slices, minced garlic, sliced onion, and chopped scallion. Stir-fry this mixture for about 1 minute over high heat, then reduce the heat to medium-low.

4. Add the soy sauce and organic sugar to the pan, stirring and mixing well with the other ingredients. Sprinkle in the sesame seeds and drizzle the roasted sesame oil. Continue to stir and mix until everything is well combined. Serve your delicious eggplant side dish.

 Note

If you prefer an alternative method, you can steam the eggplant slices, let them cool, and then season them with soup soy sauce, garlic, sesame seeds, and sesame oil.

Stir-fried Eggplant & Anchovy (Gaji Myeolchi Bokkeum / 가지멸치볶음)

Total Time: 30 minutes

Ingredients (for 4 persons):
- 2 medium-sized eggplants (12 ounces)
- 2 ounces of anchovy, medium-sized (use them as they are)
- 3 tablespoons of grapeseed oil
- 3 cloves of garlic, sliced
- ½ sweet onion, chopped (¼ cup)
- 1 small red pepper, sliced sideways (½ ounce)
- 1 small green pepper, sliced sideways (½ ounce)
- 1 ½ tablespoons of soy sauce
- 1 ½ tablespoons of red pepper powder
- 1 tablespoon of organic sugar
- 1 tablespoon of mirin
- 1 tablespoon of water
- 1 teaspoon of roasted sesame seeds

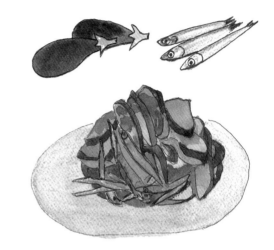

Method:

1. Begin by cutting the cleaned eggplants in half lengthwise, and then slice them into ⅛-inch-thick pieces. Set them aside.

2. Heat a frying pan over high heat and add a generous amount of oil. Once the oil is hot, start by stir-frying the sliced garlic, then quickly stir-fry the anchovy. Pour in the sugar, soy sauce, red pepper powder, and mirin, and bring the mixture to a boil.

3. Add the eggplant and chopped onion to the pan, cover with a lid for 1 minute, then open the lid and add the green pepper, red pepper, and sesame seeds. Pour in some water to help soften the eggplant flesh.

4. Serve this flavorful dish.

Tip:

Eggplants easily become soggy, but with anchovies, the flavor is enhanced, and the sogginess disappears.

Stir-fried Radish (Mu Namul / 무나물)

Total Time: 30 minutes

Ingredients (for 2 persons):
- ½ small Korean radish (approximately 1 cup)
- 1 tablespoon of grapeseed oil
- 1 clove of garlic, minced
- ½ stalk of scallion, chopped
- 1 teaspoon of Kosher salt
- ⅛ teaspoon of sliced red pepper
- ½ teaspoon of roasted crushed sesame seeds
- ½ teaspoon of roasted sesame oil

Method:

1. Begin by cutting the radish lengthwise in half. Place one half flat on your cutting board and slice it crosswise into ⅛-inch thickness. Be mindful not to make the slices too thin.

2. Heat a thick stir-frying pan and swirl the grapeseed oil around. Once the oil is hot, add the radish slices, minced garlic, and chopped scallion. Stir-fry this mixture for about 1 minute over high heat. Then, reduce the heat to medium and season it with salt and sliced red pepper. Continue to stir and mix until the radish slices lose their rawness.

3. Toss in the roasted, crushed sesame seeds and roasted sesame oil.

4. Serve this flavorful stir-fried radish dish as a side with your main course.

 Notes
- Stir-fried vegetables are not only easy to prepare but also provide essential fatty acids and enhance the taste of Korean cuisine.
- You can use various vegetables for stir-frying, such as zucchini, garlic shoots, eggplant, sesame leaves, green pepper, soaked bellflower root, soaked bracken, and butterbur stalks.

Stir-fried Red Pepper Paste (Bokkeum Gochujang / 볶음고추장)

Total Time: 30 minutes

Ingredients (for 4 persons):
- 5 ounces of beef chopped and diced
- ½ tablespoon of soy sauce
- 2 tablespoons of organic sugar
- 1 tablespoon of mirin
- 3 cloves of garlic, minced
- sprinkle of pepper
- 6 tablespoons of red pepper paste
- ½ sweet onion, chopped
- ½ tablespoon of grapeseed oil
- 1 tablespoon of roasted sesame seeds
- ½ tablespoon of roasted sesame oil

Method:

1. Start by chopping and dicing the beef. Season the beef with ½ tablespoon of soy sauce, ½ tablespoon of sugar, mirin, a sprinkle of pepper, 1 teaspoon of minced garlic, and more pepper to taste. Allow it to marinate for 5 minutes.

2. Heat a frying pan with grapeseed oil until hot. Stir-fry the marinated beef in the pan, then add the chopped sweet onion and red pepper paste.

Mix everything together thoroughly and continue stir-frying.

3. Add the remaining sugar, mirin, garlic, and more pepper to the pan. Keep stirring and frying until the beef and onions are cooked and well coated with the sauce.

4. Finish the dish by sprinkling roasted sesame seeds and drizzling roasted sesame oil over the top.

Stir-fried Soybean Sprouts (Kongnamul Bokkeum / 콩나물볶음)

Total Time: 20 minutes

Ingredients (for 4 persons):
- 1 package of soybean sprouts (12 ounces)
- 2 cloves of garlic, minced
- ½ stalks of big green onion, chopped
- 1 ounce of red pepper, chopped
- 1 ½ tablespoons of grapeseed oil
- 2 teaspoons of Kosher salt
- 1 teaspoon of soup soy sauce
- 1 teaspoon of roasted sesame seeds
- 1 teaspoon of roasted sesame oil

Method:

1. Begin by trimming and thoroughly washing the soybean sprouts. Ensure they are well-drained and set them aside.

2. Heat a frying pan over high heat and add a generous amount of oil. Once the oil is hot, carefully add the soybean sprouts to the pan. Stir in the minced garlic, chopped big green onion, red pepper, salt, and soup soy sauce.

3. As you cook, you'll notice a delightful savory aroma emanating from the soybean sprouts. This dish has a distinct flavor compared to traditional soybean sprout muchim.

4. Finish by sprinkling roasted sesame seeds and drizzling roasted sesame oil over the dish. Serve and enjoy your savory soybean sprouts.

Stir-fried Zucchini (Hobak Bokkeum / 호박볶음)

Total Time: 20 minutes

Ingredients (for 4 persons):
- 2 medium-sized zucchinis
- ½ tablespoon of grapeseed oil
- 2 cloves of garlic, minced
- 1 stalk of scallions, chopped
- 2 teaspoons of shrimp sauce
- 1 red pepper, sliced (1 ounce)
- 1 teaspoon of roasted sesame seeds
- 1 teaspoon of roasted sesame oil

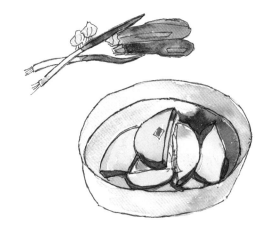

Method

1. Cut the zucchinis in half lengthwise, lay them flat, and slice them into ¼-inch-wide pieces. Try not to make them too thin.

2. Heat the grapeseed oil in a thick frying pan. When the oil is hot, add the zucchini, garlic, and scallions. Stir-fry for 1 minute over high heat, then add the shrimp sauce and red pepper. Continue to stir-fry for an additional 2 minutes over medium heat. When the zucchini becomes tender, add the sesame seeds and sesame oil. Serve.

Tip:

You can also enhance this dish by adding small dry shrimps, anchovies, peanuts, or walnut pieces while stir-frying the zucchini.

Vinegary Cucumber Side Dish (Oi Chomuchim / 오이초무침)

Total Time: 20 minutes

Ingredients (for 4 persons):
- 1 English cucumber thinly sliced
- ½ teaspoon of Kosher salt for rubbing
- ½ teaspoon of Kosher salt for the marinade
- 1 tablespoon of white vinegar
- 2 tablespoons of organic sugar
- 1 tablespoon of red pepper powder
- 1 tablespoon of chopped big green onion
- 1 tablespoon of roasted and crushed sesame seeds

Method:

1. Start by washing the cucumber. Rub the surface with ½ teaspoon of salt, then cut it in half lengthwise. Proceed to slice each half into thin ⅛ -inch pieces.
2. Prepare the marinade by mixing ½ teaspoon of salt, vinegar, and sugar. Stir until the sugar completely dissolves.
3. Thoroughly coat the cucumber slices with the marinade created in step 2. Add the red pepper powder, chopped big green onion, and sesame seeds, and mix them in.

 Note

In this recipe, there's no need to use garlic. Enjoy the simple, refreshing taste of the cucumber!

Vinegary Lotus Root (Yeongeun Chojeolim / 연근초절임)

Total Time: 20 minutes

Ingredients (for 6 persons):

- 1 row of lotus roots (12 ounces)
- 2 tablespoons of white vinegar (for soaking)
- 1 cup of white vinegar (for the vinegary liquid)
- 1 tablespoon of Kosher salt
- ¼ beets (2 ounces)
- ¾ cup of organic sugar
- ⅓ teaspoon of salt

Method:

1. Start by cleaning the lotus root. Use a potato peeler to peel off the skin and cut off both tips.
2. Cut the lotus root into very thin strips, about ⅛ -inch thick. Thinner strips work best for this recipe.
3. Place the lotus root strips in water with 2 tablespoons of vinegar. This will prevent them from turning brown and help remove any bitter taste.
4. In a pot of boiling water, add the Kosher salt. Blanch the lotus root strips for 1 to 2 minutes, then immediately rinse them with cold running water and drain.
5. Take a container with a lid and place the drained lotus root slices at the bottom. Add a chunk of beets as well.
6. In a separate pot, combine 1 cup of white vinegar, ¾ cup of organic sugar, and ⅓ teaspoon of salt. Bring this mixture to a boil.
7. Pour the hot vinegar-sugar mixture over the lotus root slices and beets in the container. Allow the mixture to cool.
8. Once cooled, cover the container with the lid and refrigerate it. After a day, you can enjoy these crispy, vinegary, and sweet lotus root slices with a lovely light carmine color.

These pickled lotus root slices make for a delightful and visually appealing side dish.

White Radish Pickle (Tongdak Jipmu / 통닭집무)

Total Time: 20 minutes

Ingredients (for 4 persons):
- ½ small Korean radish
- 4 teaspoons of Kosher salt
- 1 cup of water
- ½ cup of white vinegar
- ⅔ cup of white sugar

 Note

The standard ratio for the sweet and sour mixed water is water : vinegar : sugar=2 : 1 : 1, or it can be less. The sugar: salt ratio is 10 : 1

Method:

1. Cut the clean radish into ½-inch cubes and place them in a mason jar or any container with a lid.

2. Mix water, vinegar, and sugar in a pot and bring the mixture to a boil. Ensure that the sugar completely dissolves. Once it has boiled, pour it over the radish cubes.

3. Allow the mixture to cool, then cover it with a lid and store it in the refrigerator. After one day, the flavor of the radish cubes will develop enough to be served.

Tip:

You can make pickles with onion, celery, cabbage, or cucumber using the same method.

Chapter III

Rice, Porridge & Noodles

Abalone Porridge (Jeonbokjuk / 전복죽)

Total Time: 30 minutes

Ingredients (for 4 persons):

- 6 abalone, 10~11 mi (1 pound)
- 2 cups of rice
- 4 tablespoons of roasted sesame oil
- 1 teaspoon of Kosher salt
- 3 quarts of water

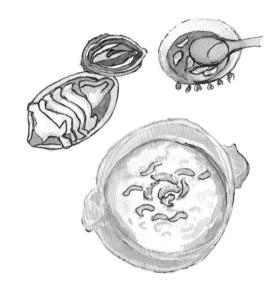

Method:

1. Allow the rice to swell in cold water for more than 1 hour. Then, strain the water through a sieve and drain the rice. Set it aside.
2. Treating the abalone: Clean the body and the shell by rubbing them under cold running water using a brush.
3. Place the cleansed whole abalone on a cutting board and separate the body part from the shell with a flat spoon or knife, pushing and turning the spoon or knife toward the center. Set it aside.
4. Slice the flesh part thinly, remove the abalone gut, and grind it in a blender while adding ½ cup of water.
5. Heat a pot and swirl around the sesame oil. Stir-fry the sliced abalone flesh and the ground gut.

6. Add the drained rice (step 1) and keep stir-frying until the rice grains turn transparent. Pour in 3 quarts (12 cups) of water and continue boiling over medium heat. Stir the ongoing porridge to the bottom of the pot with a wooden spatula to prevent scorching. Season with Kosher salt.

This porridge will have a subtle green color with a savory flavor. Many people believe that abalone is a powerful health enhancer, especially in abalone porridge.

Beef Rice Bowl (Gogi Deopbap / 고기덮밥)

Total Time: 30 minutes

Ingredients (for 2 persons):
- ½ cup of shredded beef
- 1 medium onion, thickly sliced
- 3 ounces of oyster mushrooms (or raw shiitake mushrooms), sliced
- 2 egg yolks
- ¼ stalk of big green onion, chopped
- ⅛ teaspoon of sesame seed
- 1 ½ cups of water
- 1 sheet of kelp (4 inches x 4 inches)
- 1 fistful of katsuobushi
- 3 tablespoons of soy sauce
- 1 tablespoon of mirin
- 1 tablespoon of organic sugar
- 1 ½ cloves of garlic
- sprinkle of pepper
- 4 ½ to 5 ounces of cooked rice

Method:

1. In a medium pot, bring water and kelp to a boil and simmer for about 5 minutes. Add katsuobushi and let it stand for another 5 minutes. Strain through a basket layered with a thin cotton cloth to create the broth.

2. In a frying pan, pour the katsuobushi broth from step 1, and cook the thick onion slices in it until they become transparent. Add shredded beef and mushroom slices and cook for about 2-3 minutes. Be careful not to overcook the beef; it should remain tender and flavorful.

3. In individual bowls, place cooked warm rice first, and then add the beef, onion, and mushrooms. Pour ⅔ to ¾ cup of the prepared broth over the ingredients. In the center on top, place the egg yolk, and sprinkle it with sesame seeds and chopped big green onion.

4. Serve your delicious beef and mushroom rice bowl immediately.

Bibimbap (비빔밥)

Total Time: 1 hour

Ingredients (for 1 person):
- 4 ½ ounces of steamed rice
- 2 ounces of beef, sliced and seasoned with soy sauce, garlic, sesame seed, and sesame oil, then stir-fried
- 2 ounces of soybean sprout dish
- 2 ounces of parboiled spinach dish
- 2 ounces of bracken dish
- 2 ounces of braised bellflower roots (or substitute with radish or carrot dish)
- 2 ounces of stir-fried zucchini dish
- 1 egg, fried in a pan
- 1 tablespoon of red pepper paste, seasoned with organic sugar, sesame oil, sesame seeds, and water.

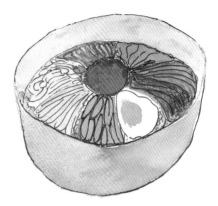

Method:

1. Take an individual serving bowl and spread out the steamed rice evenly at the bottom.
2. Arrange the various side dishes in a circle on top of the rice. You should have stir-fried beef, five different vegetable dishes, and a fried egg. Leave the center of the bowl empty.
3. In the center of the bowl, place the seasoned red pepper paste.
4. Serve your bibimbap as is, and feel free to enjoy it with a side of soybean paste soup.

Bibimbap is a versatile and customizable dish, so you can adjust the ingredients and seasonings according to your preferences.

Black Bean Sauce Noodles (Jjajangmyeon / 짜장면)

Total Time: 1 hour

IIngredients (for 3 persons):

- 8 ounces of Chinese noodles
- 1 ½ ~2 tablespoons of fermented black bean paste
- 1 ¼ tablespoons of grapeseed oil
- 6 ounces of pork
- 1 cup of chopped cabbage
- ¾ cup of chopped onion
- 1 medium carrot, chopped
- 1 stalk of celery, chopped
- ½ of zucchini
  (seasonal choice; potato, Napa cabbage)
- 1 ½ teaspoons of organic sugar
- 1 tablespoon of cornstarch
- 4 tablespoons of water

Method:

1. Cut the pork and all the vegetables into sizes between chopping and dicing, separating each to distinguish them.
2. Heat a wok, add oil, and stir-fry the pork over high heat until the surface turns brown. Avoid rushing, as it can create unpleasant bubbles in your black bean sauce.
3. Add the chopped cabbage and stir-fry until all the cabbage leaves become soft and transparent. Then, follow the same process with the onion cuts until they become transparent. Continue with the carrot, celery, and zucchini.
4. Once the zucchini is cooked, add the black

bean paste and stir well to combine all the ingredients with the paste. Ensure everything is thoroughly mixed before adding the sugar.

5. Pour in the water and boil the mixture while stirring to ensure it doesn't stick to the bottom of the wok. Add the cornstarch-water mixture and boil for about 15 seconds.

6. In a wide and deep pot, fill it with water to ¾ depth and bring it to a boil. Add the Chinese noodles and, as soon as they start boiling, pour in 1 cup of cold water. Repeat this process with an additional cup of cold water. Once it boils again, wash the cooked noodles in cold running water and drain. Make the drained noodles hot by immersing them in very hot water, drain them in a basket, and serve with plenty of the black bean sauce you prepared earlier. Top it with cucumber slices.

Tip:

The amount of black bean paste is minimal. Using too much can make the dish overly salty. Aim for approximately ½ to ⅔ tablespoon of fermented black bean paste per person.

Chinese Udong (Junghwa Udong / 중화우동)

Total Time: 45 minutes

Ingredients (for 2 persons):
- 13 ounces of Chinese noodles
- 3 ounces of pork, sliced
- 1 ½ tablespoons of grapeseed oil
- 4 medium shrimps
- ¼ cup of cuttlefish
- ⅛ cup of oyster meat (optional)
- ¼ cup of raw shiitake mushrooms
- 2 sheets of wood ear mushrooms, swollen and cut in half (optional)
- half of sweet onion, sliced
- 3 leaves of Napa cabbage, cut in half lengthwise and then into 2-inch width
- 3 cloves of garlic, minced
- ¼ teaspoon of pepper
- 1 tablespoon of Kosher salt
- ½ white stalk of big green onion
- 1 teaspoon of chicken powder (optional)
- 2 eggs, beaten
- 5 cups of water
- one head of Bok Choy with removed roots (optional)
- 1/10 bundle of chives, cut into 2-inch pieces (2 ounces)

Method:

1. Heat a wok over high heat, pour in the oil. Once the oil is hot and starts to boil, add the sliced pork, and stir-fry it. Add all the seafood - cuttlefish, shrimp, and oyster meat - and stir-fry them together for about 2-3 minutes. Then, add the prepared vegetables along with garlic, pepper, and salt. Set this mixture aside.

2. Prepare the Chinese noodles. In a large pot, bring water to a boil, then add 1 cup of cold water to the boiling water twice. After about 5 minutes, remove the noodles, rinse them with

cold water, and drain. Set them aside.

3. Beat the eggs well.

4. In another medium pot, bring water to a boil, add chicken powder, and season with salt. Let the beaten eggs flow into the boiling water, gently swirling them with chopsticks to create a soft mixture. Mix this soup with the mixture from step 1. Pour it over the noodles in individual bowls and add Bok choy and chives on top.

5. Serve and enjoy your delicious noodle soup.

Curry Rice (카레라이스)

Total Time: 30 minutes

Ingredients (for 4 persons):
- 6 ounces of pork (beef or spam), chopped
- 1 ½ tablespoons of salted butter
- ½ medium potato, chopped
- ½ medium sweet potato, chopped
- 1 medium sweet onion, chopped
- ½ medium carrot, chopped
- ½ medium zucchini, chopped
- ½ cup of cabbage, chopped
- 1 stalk of celery, diced
- 2 cups of water
- 3 ½ ounces of curry
  (4 squares, S&B Golden Curry)
- 4 ½ ~5 ounces of cooked rice on each plate

Method:

1. Chop all the ingredients, except for the celery. Dice the celery.

2. In a large pot or pan, melt the butter over high heat. Stir-fry the pork (beef or spam) until it starts to brown.

3. Add the potatoes and sweet potatoes and continue stir-frying until the pieces become half transparent.

4. Stir in the sweet onion, carrot, zucchini, cabbage, and celery. Keep stir-frying until the vegetables start to soften.

5. Pour in 2 cups of water. Add the curry paste (4 squares) to the mixture. You can adjust the amount of water to your preference for the thickness of the curry sauce, typically between 1.5 to 2 cups. Stir and cook until the curry paste is fully dissolved, and the sauce thickens.

6. Serve the curry over cooked rice and enjoy your homemade curry rice with your favorite vegetables.

Eggplant Rice Bowl (Gaji Deopbap / 가지덮밥)

Total Time: 30 minutes

Ingredients (for 3 persons):

- 2 cups of rice
- 1 ¾ cups of water
- 2 medium-sized eggplants, sliced into ⅛-inch-thick half-moon shapes
- 2 ounces of ground pork (80% lean)
- ¼ stalk of big green onion, chopped
- 1 clove of garlic
- 1 ½ tablespoons of grapeseed oil
- 2 tablespoons of soy sauce
- 1 teaspoon of mirin

Method:

1. Rinse the rice in water and let it soak for about 30 minutes. Afterward, drain and set it aside.
2. Slice the eggplants into ⅛-inch-thick half-moon shapes. Set them aside.
3. Heat a pan over high heat. Once the pan is hot, add the grapeseed oil and let it come to a boil. Reduce the heat to medium, toss in the chopped big green onion, and stir-fry. Add the ground pork and garlic, then add the sliced eggplants along with the soy sauce and mirin. Continue stir-frying.

 Note

Now it is time to make rice. A stone pot is best. Nonetheless, any other thick pot is okay to make good rice.

4. In a pot, combine the soaked and drained rice with water. Add the eggplant mixture from step 3 on top. Over medium heat, start making the rice, covering the pot with a lid for about 15 minutes. Reduce the heat to the minimum setting and let it simmer for an additional 10 minutes. Turn off the heat and allow it to stand for 5 minutes. Serve the dish with kimchi.

Gimbap (김밥)

Total Time: 1 hour

Ingredients (for 4 persons):
- 2 ½ cups of rice (seasoned with ½ tablespoon of salt, 1 tablespoon of sesame oil, and ½ tablespoon of sesame seeds)
- 10 eggs
- 4 ounces of beef flank steak
- 1 ½ sheets of fishcake, rectangular shape (4 ounces)
- ½ bunch of spinach
- ½ root of burdock
- ½ medium carrot
- ¼ medium pickled radish
- 7 ½ sheets of gim
- ½-inch piece of ginger

Method:

1. After cooking the rice, transfer it to a large bowl. Sprinkle with salt, sesame oil, and sesame seeds, and gently mix. Avoid mixing the rice roughly, as it can clump together and not spread evenly. Let the rice mixture cool for 5-10 minutes.

2. In a large deep bowl, break the eggs and whisk them. Use a square frying pan to make thin egg sheets over low heat. Cut each sheet into three pieces. These will be used as "egg blankets."

3. Slice the beef flank steak thickly and season it with soy sauce, sugar, garlic, pepper, and mirin. Stir-fry the seasoned beef in a pan. The ratio of soy sauce to sugar should be 1:1 (similar ratio for the fishcake and burdock).

4. Parboil the fishcake, then wash, slice, and braise it with soy sauce, sugar, ginger, and mirin.

5. Boil the burdock slices for 30 seconds, then wash and drain. Braise them in soy sauce and sugar.

6. Slice the carrot and parboil it in boiling water, then drain.

7. Blanch the spinach, wash it in cold water, and drain. Squeeze it and season with salt.

8. Slice the pickled radish a little thicker than the other ingredients.

9. Place rice evenly on 1.5 sheets of gim/laver on a bamboo sheet. Place an egg blanket in the middle.

10. On the center of the egg blanket, start by placing spinach and fishcake. Then add all the slices (beef, carrot, burdock, and pickled radish). Roll the egg blanket toward you, then roll the gim/laver and bamboo sheet together toward yourself. Press the bamboo sheet firmly once or twice. Brush sesame oil on one side of the gimbap.

11. Cut into 0.5 to 0.7-inch slices. It's both visually pleasing and delicious! Serve and enjoy.

I prefer using the spelling 'gim' over 'kim' as it closely mirrors the original sound.

Knife-cut Noodle Soup with Seafood (Kalguksu / 칼국수)

Total Time: 30 minutes

Ingredients (for 2 persons):
- 11 ounces of knife-cut noodle
- 6 large dried anchovies
- 2 large-eyed herrings
- ½ palm size of kelp
- ½ tablespoon of Kosher salt
- 3 ounces of clam flesh
- 2 shrimps
- 1 medium zucchini, chopped
- ¼ cup of cabbage, chopped
- ¼ medium carrot, chopped
- ½ onion, chopped
- 1 stalk of scallion, chopped
- 3 cloves of garlic, sliced
- ⅛ -inch piece of ginger
- ¼ teaspoon of pepper
- 5 cups of water

Method:

1. In a pot, bring 5 cups of water to a boil. Add large dried anchovies, large-eyed herrings, and kelp. Boil the contents for about 7-8 minutes, then remove the anchovies, large-eyed herrings, and kelp. This creates anchovy broth.

2. While the anchovy broth is boiling in step 1, prepare the vegetables by chopping the zucchini, cabbage, carrot, onion, and scallion. Slice the garlic and ginger thinly. Remove any excess powder from the noodles by shaking them and place the noodles on a large plate. Set it aside.

3. Add the chopped vegetables, scallion, and salt to the anchovy broth from step 1 and bring it to a boil. Add the knife-cut noodles and continue boiling for about 5 minutes over medium heat.

4. Increase the heat to high and add the clam flesh and shrimp. Cook for an additional 3 minutes. Adjust the seasoning with scallion and

salt to taste (optional). Serve.

 Note

Choosing the right kind of knife-cut noodle is
crucial for making a good noodle soup. In the
past, it was common for families to make dough
by kneading, rolling it out on a flat table with
flour, and cutting it with knives. These days,
such practices are rare, but it's a tradition worth
preserving.

Korean Ginseng Chicken Soup (Samgyetang / 삼계탕)

Total Time: 1 hour

Ingredients (for 2 persons):

- 1 small Cornish hen (about 3 pounds)
- 1 ½ cups of glutinous rice
- 2 roots of fresh ginseng
- 12 cloves of garlic
- 4 jujubes
- ¼-inch piece of ginger
- 1 big onion
- 1 stalk of celery
- ½ tablespoon of Kosher salt
- 2 teaspoons of mirin
- 1 ½ quarts of water

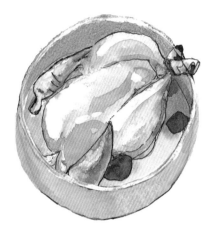

Method:

1. Wash and clean the glutinous rice, soak it in cold water for at least 30 minutes, and then drain. Set it aside.

2. Remove all giblets from the chicken cavity and wash the whole chicken thoroughly, both inside and outside, under running cold water. Drain and set aside.

3. In a large pot, prepare 1.5 quarts (6 cups) of water and add the salt.

4. Stuff the chicken cavity with one-third of the soaked glutinous rice, ginseng roots, ginger, one-third of the garlic, and mirin. Place the stuffed chicken on the bottom of the pot.

5. Add the remaining two-thirds of the glutinous rice, the rest of the garlic, jujube, onion, celery, and water to the pot.

6. Bring the contents to a boil over high heat for about 30 minutes, then reduce the heat to medium and simmer for approximately 1 hour. To check if it's done, gently shake one of the chicken's legs. If it moves easily, your Samgyetang is ready to be served.

This traditional Korean chicken soup is both delicious and nutritious. Enjoy your Samgyetang!

Mixed Cold Buckwheat Noodles (Bibim Naengmyeon / 비빔냉면)

Total Time: 40 minutes

Ingredients (for 4 persons):
- 1 ⅓ pounds of dry buckwheat noodles
- 1 large pear, sliced
- 2 medium cucumbers, sliced
- 6 tablespoons of hot pepper paste
- 3 tablespoons of soy sauce
- 3 tablespoons of red pepper powder
- 3 tablespoons of organic sugar
- 3 tablespoons of rice vinegar
- 3 tablespoons of plum extract
- 4 stalks of chopped and diced scallion
- 2 cloves of garlic, diced
- 1 medium-sized onion, chopped and diced
- 2 teaspoons of roasted and crushed sesame seeds
- 1 tablespoon of roasted sesame oil (add just before serving)

Method:

1. Slice the pear and cucumber, ensuring the slices are not too thin. Set them aside.

2. Mix all the ingredients together in a bowl, except for the sesame seeds and sesame oil. Set this marinade mixture aside.

3. Boil water in a large pot. Add the dry buckwheat noodles and boil for 4 minutes. Rinse the noodles under fast-running cold water, rubbing them vigorously to remove any starch powder that may be coated on them by the manufacturer. Ensure the water is cold. Drain the noodles using a basket.

4. In individual bowls, start by placing the noodles, then add the marinade mixture from step 2. Finally, arrange the pear and cucumber slices on top of everything. Garnish with hard-boiled egg halves, sesame seeds, and drizzle roasted sesame oil just before serving.

Multi-grain Rice (Japgokbap / 잡곡밥)

Total Time: 1 hour

Ingredients (for 10 persons):
- ⅓ cup of white rice
- ⅓ cup of whole grain rice
- ⅓ cup of whole grain glutinous rice
  (You can adjust the ratio of these three grains)
- ⅓ cup of whole grain oat
- ⅓ cup of pearl barley
- ⅓ cup of glutinous sugarcane
- ⅓ cup of Job's tears
- ⅓ cup of glutinous millet
- 1/12 cup of black rice (a little less)
- ¼ cup of green flesh black beans
  (a little more)
- 4 ½ cups of water (more than regular rice)

Method:

1. Wash the first and second grain-washed water to remove any dust and debris that may be present with the grains. Wash them 2-3 times more for thorough cleaning.

2. Immediately after washing and cleaning the grains, set your electric pressure rice cooker to the multi-grain setting.

3. After the rice is cooked, wait for 3-5 minutes to allow the rice pot to let the grain contents soften.

4. Mix the multi-grain rice thoroughly from bottom to top using a rice scoop. Serve.

 Notes

• Carbohydrates are an essential source of energy for our daily life. It's crucial not to misunderstand or consider carbohydrates harmful or toxic to our bodies. Misinformation and misconceptions about nutrition can lead to an imbalanced diet. When carbohydrates aren't adequately supplied, our bodies search for alternative energy sources, such as muscle protein, which can result in muscle loss and premature aging.

• We do not have the enzymes to break down the cellulose found in the outer shells of multi-grains. These enzymes were not naturally provided to us. A pressure cooker can help soften these multi-grains, allowing them to pass through our bodies. These grains offer various health benefits, including slowing the absorption of glucose, enhancing intestinal movement, and providing a feeling of fullness. While some recent research may suggest that whole grains have downsides, they still offer significant health benefits.

• It's advisable not to mix the grains together in your pantry, as each grain metabolizes at a different rate. Mixing them together can cause the heat generated by individual grains to affect one another, potentially leading to spoilage. Label each container and store them in the refrigerator to prevent insect infestations during the hot seasons.

New Year's Day Rice Cake Soup (Tteokguk / 떡국)

Total Time: 30 minutes

Ingredients (for 4 persons):
- 1.8 pounds of sliced rice cake
- 1 stalk of big green onion
- 10 large dried anchovies
- 4 large-eyed herrings
- 1 piece of kelp (1 palm-sized piece)
- 1 ½ quarts of water
- 1 tablespoon of Kosher salt
- 3 eggs
- 6 ounces of beef, sliced into 1.5-inch lengthwise and ⅛-inch width.
- 1 clove of garlic
- 4 medium-sized shiitake mushroom (½ ounce)
- 1 tablespoon of grapeseed oil
- ½ teaspoon of roasted sesame oil
- ½ ounce of gim/laver (4 pieces, raw, roasted)
- ½ teaspoon of pepper

Method:

1. Boil water and make a broth by adding anchovies, large-eyed herring, and kelp.

2. Soak shiitake mushrooms and rice cakes in water to soften them. Drain and set aside.

3. Season the beef slices with salt and roasted sesame oil, then stir-fry them in a frying pan. Set it aside.

4. Separate the egg whites and yolks. Add salt to each and whisk to make shredded white and yellow egg crepes.

5. Thinly slice the rehydrated shiitake mushrooms and stir-fry them with salt, pepper, and sesame oil.

6. Break the raw gim (laver) into small pieces.

7. In the broth from step 1, add the rice cakes and boil for about 20 minutes. Then add big green onions and salt. Serve.

8. Prepare the 5 garnishes separately: beef, yellow egg crepe, white egg crepe, mushroom, and laver.

To serve: Each person can add their choice of garnishes from the selection on the dining table to their bowl of rice cake soup. Enjoy!

 Note

While many families use bone soup or brisket broth, this recipe uses anchovy broth, a method passed down from my mother. Rice cake soup is traditionally served on New Year's Day but has become popular and is enjoyed throughout the year.

Omurice (오므라이스)

Total Time: 1 hour

Ingredients (for 4 persons):
For the Rice:

- 2 to 2 ½ cups of rice
- Water
  (less than the regular amount for rice cooking)

For the Fried Rice:

- 6 ounces of beef, pork, or ham
  (cut into medium-sized dice)
- 8 eggs
- 1 teaspoon of Kosher salt
- 1 teaspoon of mirin
- pepper (to taste)
- 1 medium-sized onion
- half a medium carrot
- 1 stalk of celery
- 1 medium-sized zucchini
- 4 raw shiitake mushrooms (¼ cup)
- 2 tablespoons of grapeseed oil
- 1 ½ tablespoons of salted butter
- ⅔ cup of tomato ketchup
  (used divided by two)
- 1 tablespoon of soy sauce
- 2 teaspoons of rice vinegar

Method:

1. Cook the rice in an electric rice cooker.
2. Cut the meat (beef, pork, or ham) into medium-sized dice and chop it. Season with salt, mirin, and pepper. Stir-fry the seasoned meat with grapeseed oil in a heated frying pan. Set it aside.
3. Chop and dice the carrot, onion, celery, zucchini, and raw shiitake mushrooms. Add the stir-fried meat to the pan and mix. Stir for 30 seconds more. Push the ingredients to one side

of the frying pan. On the other side of the pan, mix and stir together ⅓ cup of tomato ketchup, butter, soy sauce, and rice vinegar to make a sauce. Add the cooked rice and increase the heat. Mix all the ingredients together and then set them aside.

4. In a big deep mixing bowl, break 8 eggs and whisk them. If the eggs are small, you may need to use 2 more eggs.

5. Heat a wide, shallow frying pan with grapeseed oil. Reduce the heat and spread one scoop of the egg batter in the pan. When the egg batter is half cooked, add ¾ to 1 cup of fried rice on top of it. Flip it over (egg + fried rice) onto an individual serving plate. Decorate the top with ketchup.

Tip:

The egg batter should not be too thick or burnt. You might need some practice to get the flipping right. A square egg frying pan is recommended for flipping the egg and rice together, as it's light and allows your plate to be close to the fried rice and the egg batter underneath. It may take some trial and error to master this technique.

Pesto Sandwiches

Total Time: 10 minutes

Ingredients (for 1 person):
- 2 slices of bread
- 3~4 leaves of romaine lettuce
- ⅛ medium sized sweet onion, thinly sliced
- ⅛ medium sized tomato, thinly sliced
- ¼ avocado, sliced
- ¼ lemon, juiced by squeezing
- 1 sheet of cheddar cheese
- 1 sheet of white cheese
- 2 slices of smoked ham
- 2 slices of turkey breast
- 3 slices of roast beef
- 1 tablespoon of pesto

Method:

1. Wash the romaine lettuce and drain. Set it aside.

2. Slice the sweet onion and tomato thinly.

3. Slice the avocado and sprinkle lemon juice on it by squeezing. This helps prevent browning.

4. Spread pesto all over the inner walls of two slices of bread.

5. Assemble the sandwich as follows: On one slice of bread, layer lettuce, onion, tomato, avocado with lemon juice, two types of cheese, and three cold cuts in any order you prefer. Cover with the other slice of bread.

6. Cut the sandwich in half and serve.

Pine Nut Porridge (Jatjuk / 잣죽)

**Total Time: 1 hour for soaking and
30 minutes for cooking**

Ingredients (for 2 persons):
- 1 cup of rice, soaked for 1 hour
- 1 cup of pine nuts
- 2 cups of water

Method:

1. In a blender, grind the soaked rice and pine nuts with 1 cup of water. Strain the mixture into a deep pot for cooking. Add an additional 1 cup of water to the blender to collect any remaining rice, then strain it again.

2. Bring the strained liquid in the pot from step 1 to a boil. Initially, the porridge may seem too thick, but do not add any more water. It will thin out as it cooks. Serve.

Tips:

- Do not add salt during cooking so that individuals can salt their porridge to taste at the dining table.

- Keep in mind that you should use an equal amount of pine nuts as rice and twice as much water. No additional water is needed.

Rice with kimchi & Soybean Sprouts
(Kimchi Kongnamulbap / 김치콩나물밥)

Total Time: 45 minutes

Ingredients (for 4 persons):
- 2 cups of rice
- 1 ¾ cups of water
- 8 ounces of sliced pork
- ½ cut of whole kimchi, cut, and squeezed (1 cup)
- ¾ of a 12-ounce package of soybean sprouts
- 3 tablespoons of soy sauce (1 for pork, 2 for seasoning sauce)
- 1 teaspoon of ginger juice
- 2 cloves of garlic (1 for pork, 1 for seasoning sauce)
- 2 teaspoons of chopped big green onion
- 1 ½ teaspoons of pepper (0.5 for pork, 1 for seasoning sauce)
- 3 tablespoons of roasted sesame oil (1 for rice, 1 for pork, 1 for seasoning sauce)
- 1 tablespoon of roasted and crushed sesame seeds

Method:

1. Season the sliced pork with soy sauce, ginger juice, garlic, pepper, and sesame oil. Set it aside.
2. Wash the rice and soybean sprouts, then set them aside.

3. Remove the stuffing from the kimchi, cut it into small pieces, squeeze it, and mix it with the seasoned pork from step 1.
4. In a medium pot, alternate layers of rice from step 2 and the mixture from step 3. Pour in 1 and ¾ cups of water, cover with a lid, and start cooking the rice. When the rice mixture begins to boil, lower the heat. Add the soybean sprouts when the water simmers down and continue to cook on very low heat until the soybean sprouts

are done and have a rich aroma. Add sesame oil and mix well before serving.

Seasoning Sauce for Rice with Kimchi and Soybean Sprouts:

- 2 tablespoons of soy sauce
- 1 teaspoon of garlic
- 2 tablespoons of chopped big green onion.
- 1 teaspoon of red pepper powder
- 1 tablespoon of sesame oil
- 1 tablespoon of roasted and crushed sesame seeds

Prepare the seasoning sauce on the dining table for diners to use as desired.

Rice with Soybean Sprouts (Kongnamulbap / 콩나물밥)

Total Time: 45 minutes

Ingredients (for 3 persons):

- 2 cups of rice
- 2 packages of soybean sprouts
  (1 ½ pounds, 12-ounce-package)
- 6 ounces of ground pork
- 20 large dried anchovies
- 1 tablespoon of mirin
- ½ stalk of scallions
- 2 cloves of garlic
- 1 teaspoon of ginger juice
- 2 teaspoons of soy sauce
- 2 teaspoons of soup soy sauce
- 1 teaspoon of red pepper powder
- a pinch of pepper
- 1 teaspoon of grapeseed oil
- 1 tablespoon of roasted sesame oil
- 1 teaspoon of roasted and crushed sesame
  seeds

Method:

1. Wash the rice and let it soak in water for 30 minutes, then drain it using a strainer. Set it aside.

2. Clean and wash the soybean sprouts, then drain them using a strainer. Set them aside.

3. Prepare a broth by using the large dried anchovies. Let it cool and set it aside.

4. Season the ground pork with ginger juice, 1 teaspoon of soy sauce, pepper, and 1 teaspoon of garlic. Set it aside.

5. In an electric rice cooker, add the following layers: seasoned pork (step 4), anchovy broth (step 3), and soybean sprouts (step 2). Be careful not to exceed the 2-cup mark. Turn on the rice

cooker, setting it to cook nutritious rice. When it's finished cooking, let it sit for 5 minutes.

6. While waiting for the rice cooker, prepare the seasoning sauce: 2 teaspoons of soup soy sauce, 1 teaspoon of soy sauce, 2 tablespoons of scallions, 1 teaspoon of garlic, 1 teaspoon of red pepper powder, 1 tablespoon of sesame oil, and 1 teaspoon of roasted and crushed sesame seeds.

Spicy Korean Acorn Noodles (Bibim Dotori Guksu / 비빔도토리국수)

Total Time: 45 minutes

Ingredients (for 4 persons):

- 2 boiled eggs
- 1 cup of stir-fried ground beef (90% lean)
- 1 ¼ pounds of dry acorn noodles (new crops) (5 ounce/person)
- The volume of dry acorn noodles per person is equivalent to a cylinder formed by the space between the thumb and middle finger.
- ¼ of cabbage, sliced, washed, and drained. (2 cups)
- 1 ½ English cucumbers, sliced
- 1 stalk of scallion, chopped
- 2 cloves of garlic, minced
- 6 tablespoons of red pepper paste
- ¼ cup of rice vinegar
- 3 tablespoons of organic sugar
- 2 tablespoons of roasted sesame oil
- 1 tablespoon of roasted and crushed sesame seeds

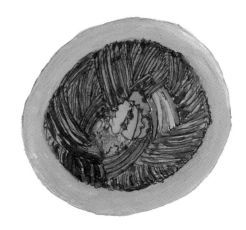

Method:

1. In a wide, large pot, fill it with water to more than half and bring it to a boil. Add the acorn noodles. When the noodles boil up with cloudy bubbles, add 1 cup of cold water. Once they boil up again, add one more cup of cold water. At the third boiling, flip over the cooked noodles into a big basket, wash them with cold running water, rubbing them three or four times, and then drain them on this basket. Set them aside.

2. Quickly mix the cabbage, cucumber, scallion, garlic, red pepper paste, vinegar, sugar, sesame oil, and sesame seeds, incorporating all the ingredients listed above.

3. Place the cooked noodles from step 1 in individual bowls, top them with the mixture from step 2, and add half a piece of hard-boiled

egg and the stir-fried ground beef that were prepared ahead of cooking. Serve.

Tip:
You can use well-fermented sliced Napa cabbage kimchi as a replacement for cabbage slices. In this case, skip using scallion and garlic, and reduce the amount of red pepper paste by half.
Place extra red pepper paste and sesame oil on the dining table for additional seasoning.

Spicy Seafood Noodle Soup (Jjamppong / 짬뽕)

Total Time: 45 minutes

Ingredients (for 2 persons):

- 13 ounces of Chinese noodles
- 2 ounces of pork (1-inch-slice, thickly sliced)
- 4 big shrimps (4 ounces)
- ¼ cup of oyster meat (optional)
- ⅜ cup of cuttlefish
- ¼ cup of clam meat (2 ounces)
- ½ of zucchini, halved and cut thickly
- 2 leaves of Napa cabbage, halved and cut into 2-inch-width pieces
- (3 ounces) ¼ of chopped green pepper
- 2 ½ ounces of bamboo shoot (½ can of 5 drained bamboo shoots) (optional)
- 2 raw shiitake mushrooms or ⅛ oz dry shiitake mushroom, cut in half
- 2 big wood ear mushrooms, swollen, cut in half (optional)
- 3 tablespoons of grapeseed oil
- ¼ white stalk of big green onion, 1-inch cut
- 3 cloves of garlic, diced
- ½ of onion, thickly sliced
- 12 leaves of Bok Choy (½ head)
- 1 ½ tablespoon of red pepper powder
- 1 tablespoon of oyster sauce

- ½ tablespoon of Kosher salt
- ½ tablespoon of soy sauce
- 1 ½ pint of beef bone soup or crucible soup (add ⅕ pint more)

If it is difficult to shop for these ingredients {oyster meat, bamboo shoot, and wood ear mushroom} in your area, simply omit them! Instead, why don't you use more clam meat & more shiitake mushrooms?

Method:

1. Prepare all the ingredients as directed above. Set them aside.
2. Stir-fry the oil, big green onion, and garlic over high heat.
3. On top of the green onion and garlic oil from step 2, stir-fry pork and green pepper. Keep stir-frying all the vegetables except soy sauce, oyster sauce, salt, and Bok choy. Reduce heat to medium and add red pepper powder with ½ cup of water. Once they are well mixed, turn up the heat.
4. Pour beef bone soup over the mixture from step 3 and add soy sauce, oyster sauce, salt, shrimp, oyster meat, cuttlefish, and clam meat. As soon as they are cooked, reduce the heat.
5. In a wide and deep pot, fill it with water to ¾ depth and bring it to a boil. Add the Chinese noodles and, as soon as they start boiling, pour in 1 cup of cold water. Repeat this process with an additional cup of cold water. Once it boils again, wash the cooked noodles in cold running water, drain, and make the drained noodles hot by immersing them in very hot water. Drain them in a basket.
6. Blanch the Bok choy while you're working on step 5.
7. Pour the soup from step 4 over the hot noodles from step 5. Place the blanched bok choy on top of everything and serve.

Additional Tip:

If you have prepared beef bone soup at home, that's the best. Alternatively, you can purchase ready-made bone soup from a Korean market. You only need a small amount of each type of seafood, so consider packing each seafood in a small zip-lock bag for future use. This way, you can easily make Spicy Seafood Noodle Soup whenever your family members request it.

Sushi Rice

Total Time: 1 hour

Ingredients (for 4 persons):
- 2 ½ cups of Nishiki pre-washed rice
- 3-inch x 1-inch of kelp (½ palm size)
- ¼ cup of mirin (1/10 of rice)
- 2 ¾ cups of water (less than regular rice)
- ½ cup of rice vinegar
- ⅖ cup of organic sugar
- ½ teaspoon of Kosher salt

Method:

1. In an electric rice cooker, combine the rice, water, kelp, and mirin. Let it sit for 2 hours. Do not reduce the time if you want flavorful sushi rice. Remove the kelp and start cooking. After the rice cooker switches off, wait for an additional 17 minutes.

2. While waiting for step 1 to finish, prepare the vinegary sweet sauce. The ratio is vinegar to sugar is 5:4.
   - Mix ½ cup vinegar and ⅖ cup sugar. Add salt to taste (approximately ½ teaspoon).
   - You'll need an amount of vinegary sauce equal to the number of rice cups divided by 4. For 2.5 cups of rice, you'll need ⅝ cup of vinegary sweet sauce. Reserve the remaining ¼ cup of the sauce to wet your rice spatula and knife.

3. If possible, prepare a wooden bowl and a wooden spatula, along with a fan. Transfer the cooked rice to the wooden bowl using a wooden spatula. Fan the rice from time to time. This will help the rice become shiny and slightly firm. Spread the rice out into a single layer using the wooden tools.

With this sushi rice, you can make various dishes, including California rolls, mixed sashimi rice, nigiri sushi, and chirashi bowl.

 Note

For detailed instructions on how to make California rolls and other variations, please visit my Korean version at www.shinkyung.wixsite.com/cookandjoy. You can use translation tools to access additional information.

Tuna Sashimi Rice Bowl (Hoedeopbap / 회덮밥)

Total Time: 1 hour

Ingredients (for 2 persons):
- 1 ½ cups of sushi rice
  (Refer to Sushi Rice Recipe in this book)
- ¾ cup of tuna (cut into sticks and then cubed)
- ½ cup of Korean radish (sliced thinly)
- ¼ cup of lettuce
- 2 cloves of garlic sliced finely
- ¼ cup of red onion, sliced thinly
- ½ pepper, sliced thinly (2 ounces)
- smelts' roe (optional)
- quail egg (optional)
- radish sprouts (optional)

For the Vinegary Red Pepper Paste:
- 3 tablespoons of red pepper paste
- 1 tablespoon of garlic, minced.
- 1 tablespoon of rice vinegar
- 1 tablespoon of organic sugar

Method:

1. Prepare the block of frozen tuna. Thaw it on the coldest shelf of your refrigerator. When the tuna is half frozen, cut it into stick shapes first and then into cubes. You can also include smelt roe and quail egg if you have them on hand.

2. Cook rice. Let it cool in an individual bowl by fanning.

3. In a small bowl, mix the red pepper paste, minced garlic, rice vinegar, and organic sugar. This will be served on the dining table.

4. Slice the radish thinly and immerse it in cold water with a drop of vinegar for about 30 seconds. Drain it completely in a strainer. The radish slices will become transparent and stand up, looking like they're celebrating. Set it aside.

5. Wash the lettuce and slice it into ¼-inch-thick pieces. Drain and set aside.

6. Slice the garlic very finely. This is important for the dish.

7. Slice the onion and green pepper thinly. Set it aside.

8. In each individual bowl, spread out the sushi rice.

9. On top of the rice, stack the vegetables in the following order:

  - Radish slices from step 3
  - Lettuce from step 4
  - Sliced garlic from step 5
  - Sliced onion and green pepper from step 6

- Cubed tuna from step 1

Serve the Vinegary Red Pepper Paste on the dining table Each person can add it to their bowl of Tuna Bibimbap according to their taste.

<Assorted Sashimi Rice Bowl>
The open refrigerator section of raw fish at the Korean market has several packages of assorted raw fish. Usually it contains fresh tuna, fluke, and salmon. You can make "Assorted Sashimi Rice Bowl" using a package of these raw fish.

 Notes

When preparing raw fish, ensure that the raw fish is fresh and has been properly stored. You can cut them into bite-sized strips or arrange them creatively.
Remember to maintain cleanliness and ensure the freshness of the ingredients. Adjust the quantities base on your preferences, and feel free to get creative with additional toppings or sauces.

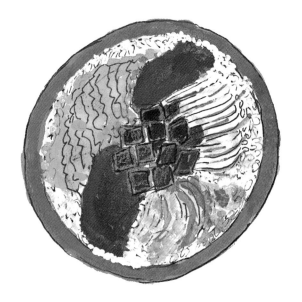

Udong (Udong / 우동)

Total Time: 30 minutes

Ingredients (for 2 persons):
- 1 sheet of kelp
  (4"x 4", big palm size, 2 sheets)
- 4 large-eyed herrings
- 10 large dried anchovies
- 1 fist of dry shrimp (⅛ ounce)
- 2 packages of udong noodles (15 ounces)
- 1 cup of katsuobushi (dried tuna)
- 3 tablespoons of tsuyu
- 1 tablespoon of soup soy sauce
- ½ tablespoon of salt
- 1 tablespoon of mirin
- 2 stalks of scallion
- 2 sheets of fishcake (3 ounces)
- 4 pieces of fried tofu (¼ ounce)
- 1/10 bundle of crown daisy (½ ounce)

Method:

1. In a pot, boil water, and add large dried anchovies, large-eyed herrings, dry shrimp, and kelp. Keep boiling for 10 minutes. Remove the kelp and boil for an additional 5 minutes. Take out the two fish 15 minutes later and discard them.

2. In the same pot, add katsuobushi and let the broth sit for 10 minutes. Afterward, filter it through a wet cotton cloth. Katsuobushi can be used for rice balls.

3. To the filtered broth, add tsuyu, soup soy sauce, salt, fish cake, and udong noodles.

4. Add fried tofu and big green onion last. Decorate with crown daisy when serving.

Watery Cold Buckwheat Noodles (Mul Naengmyeon / 물냉면)

Total Time: 2 hours

Ingredients (for 4 persons):

- 1 medium-sized radish thinly sliced
- 1 medium-sized cucumber thinly sliced
- 1 tablespoon of coarse sea salt
  (for salting radish)
- ½ teaspoon of Kosher salt (¼ for salting
  cucumber + ¼ for seasoning radish)
- ⅓ cup of rice vinegar
- ⅓ cup of organic sugar
- 1 tablespoon of red pepper powder
- 1 ½ teaspoons of ginger powder

Method:

1. Rub the cucumber with salt and cut off both ends. Lay the cucumber down on a cutting board, cut it in half, and slice it very thinly. Season with ¼ tsp salt, mix with your hands, and squeeze out excess moisture.

2. Cut the radish into thirds, then cut each third into thirds again. Slice it very thinly, then salt it with coarse sea salt for 30 minutes. Wash it under running water and drain. Set it aside.

3. Season the salted radish from step 2 with vinegar, sugar, red pepper powder, and ginger powder. Allow it to ferment. Because you

sliced the cucumber and radish very thinly, fermentation will occur quickly, usually within half a day.

Nyeongmyeon Broth Making:

Ingredients (for 4 persons):

- 1 pound of brisket
- ¼ medium-sized radish (1 pound)
- 1 medium-sized onion (8 ounces)
- 1 stalk of big green onion (8 ounces)
- ⅓ -inch piece of ginger
- 9 cloves of garlic

- 1 tablespoon of pepper
- 2 sticks of licorice (½ ounce)
- 4 quarts of water
- 3 tablespoons of soup soy sauce
- 3 tablespoons of organic sugar
- 4 tablespoon of rice vinegar
- ½ teaspoon of Kosher salt

Method:

1. Boil 4.5 quarts (+0.5) of water with brisket, radish, onion, ginger, garlic, pepper, and licorice for one hour over high heat. Strain the broth using a cotton cloth to obtain a clear broth. Save the brisket. Set it aside.

2. Season the broth from step 1 with soup soy sauce, sugar, vinegar, and salt. This is the naengmyeon broth.

3. Thinly slice the brisket from step 1. On top of cold noodles with broth, place slices of brisket, cucumber kimchi, radish kimchi, and half of a boiled egg. You can also add a slice of Korean pear!

Preparing the noodles:

- 1.3 LB of dry buckwheat noodles (Chungsoo product)

Boil water in a large pot. Add the dry buckwheat noodles and boil for 4 minutes. Wash the noodles under cold, fast-running water, rubbing them vigorously to remove the starch powder that may be coated on them by the manufacturer.

Artist: Kyung Shin
Title: Harvesting
Size: 40″×30″
Medium: oil on canvas

Chapter IV

Soups & Stews

Bean Curd Stew (Kongbijijjigae / 콩비지찌개)

**Total Time: overnight for soaking
and 1 hour and 30 minutes for cooking**

Ingredients (for 4 persons):
- 1 cup of soybeans
- 6 ounces of pork rib flesh, chopped
- ½ of fully fermented whole kimchi-cut (1 cup)
- ½ stalk of chopped big green onion
- 2 cloves of garlic, diced
- 1 tablespoon of soup soy sauce ⅛ teaspoon of Kosher salt
- pepper, to taste
- 1 teaspoon of grapeseed oil
- 3 cups of water

Stew Sauce:
- 3 tablespoons soy sauce
- 1 tablespoon garlic, diced
- 1 tablespoon scallion, chopped
- 1 tablespoon each of green and red pepper
- ½ tablespoon sesame oil
- sesame seeds, for sprinkling
- red pepper powder, for sprinkling

Method:

1. Let the soybeans soak in cold clean water (3 times the amount of water) overnight. Peel off the shells of the soybeans by rubbing them with your hands. If you shake the container, the shells will float to the surface. Remove the shells and save the swollen soybeans. Drain and set aside.

2. Boil the soaked soybeans in water. Make sure there's enough water for the soybeans to be fully immersed in. Drain.

3. Grind the drained soybeans (step 2) in a blender with enough water to cover them. You want the cooked soybeans to be able to be immersed again.

4. Season the chopped pork with big green

onion, garlic, salt, and pepper. Set it aside.
5. Remove most of the fillers from the kimchi by shaking it. Chop and slice the kimchi leaves.
6. In a pot, heat the grapeseed oil and add the seasoned pork (step 4) and chopped kimchi (step 5). Stir-fry for about 5 minutes until cooked. Once they are cooked, pour in the ground soybeans (step 3) and simmer over low heat for 20 minutes. When a savory smell starts coming out from the stew, add the big green onion and garlic. Season with soup soy sauce and simmer for an additional 5 minutes.

You can serve this stew unseasoned and provide the "stew sauce" separately for individuals to season according to their taste.
Traditionally, people from Pyongando and Hamkyungdo used whole pork ribs, boiling them for several days. This procedure is quite complicated, and the bone part can make it difficult to eat, so this simplified version is adapted for convenience. Enjoy your meal!

Beef Bone Soup (Sagol Guk / 사골국)

Total Time: 7~13 hours for boiling and 30 minutes for seasoning

Ingredients (for 4 persons):

- 1 ½ pounds of marrow bone (Korean market one 1 package)
- ½ pound of brisket
- 1 teaspoon of ginger, sliced
- 3 cloves of garlic, whole
- ¾ cup of whole onion
- 1 stalk of big green onion
- 1 stalk of celery
- 10 quarts of water: 2 quarts (for blood removal) + 4 quarts (first soup) + 4 quarts (second soup) => The water will be reduced to half at the end.

Method:

1. In a thick deep pot, place the bones and pour enough water to immerse them. Boil for about 30 minutes. Discard this liquid.

2. Pour 6 times more water than the volume of bones over the bones and boil over high heat. Once it starts boiling, reduce the heat to low and simmer for about 3-6 hours. Set aside until you make the second soup.

3. Repeat the procedure from step 2 to make the second soup. Combine the first and second soups. Boil for one more hour with a large pouch containing ginger, garlic, onion, celery, and big green onion.

4. Take out the brisket meat, rinse it with cold water, and drain. Cut the brisket meat into thin slices. Set it aside.

5. Serve the brisket meat with chopped big green onion on top of a very hot bone soup. Provide salt and pepper at the dining table for seasoning to taste.

Note

It's important to avoid excessive boiling to prevent the release of too much phosphate from the bones, which can hinder calcium absorption in the body. The third soup is sufficient for making this bone soup. This Korean Beef Bone Soup is a nourishing and comforting dish that's perfect for enjoying on chilly days.

Brisket Vegetables Rice Soup (Gari Gukbap / 가리국밥)

Total Time: 1 hour 30 minutes

Ingredients (for 4 persons):

- 1 ½ pounds of brisket
- ½ of onion (¼ cup)
- 1 stalk of big green onion (8 ounces)
- 4 ounces with root + 4 ounces, chopped
- ½-inch piece of ginger
- 3 cloves of garlic
- 1 tablespoon of Kosher salt
- 1 teaspoon of soup soy sauce
- 1 sheet of kelp (4"x4")
- 1 ½ quarts of water
- 32 fluid ounces of bone soup
  (2 packs, 1 quart)
- 1 ⅓ packages of soybean sprouts (1 pound)
- 2 handfuls of dry bracken (1 ½ ounces)
- 2 eggs to make an egg roll

For the Green Pepper Seasoning (Dadaegi):

- 1 green pepper, diced (1 ounce)
- 1 tablespoon of red pepper powder
- 1 ½ cloves of garlic, diced
- 1 tablespoon of shrimp sauce

Method:

1. Boil water and cook dry bracken for about 10

minutes over high heat. Wash the bracken with cold water, changing the water until no brown tints remain. Drain and set aside. The bracken should be softened. It will become 12 ounces.
2. In a large pot, bring 2.5 quarts of water to a boil. Add a chunk of brisket, onion, the stalk of big green onion with the root, ginger, garlic, sea salt, soup soy sauce, and kelp to make a rich beef broth. This takes about an hour, with the first 20 minutes over high heat and the remaining time over medium heat.
3. Remove the brisket from the broth, drain, and cut it into thin slices. Save the broth by filtering

out the solid ingredients. Set the broth aside.

4. Wash and clean the soybean sprouts, then drain them. Set them aside.

5. Mix the broth from step 2 with the two packs of bone soup and bring it to a boil along with the softened bracken from step 1, the sliced brisket from step 3, and the soybean sprouts from step 4. Simmer this mixture for about 15 minutes.

6. While step 4 is cooking, prepare egg roll slices for garnishing the top of the soup.

7. In individual serving bowls, place steamed rice first. Pour the soup mixture (step 4) over the rice and add the egg roll slices from step 5 on top.

8. On the dining table, provide a small bowl of chopped big green onion and a bowl of green pepper seasoning (Dadaegi).

 Note

Garigukbap is a dish originating from Hamgyong Province in the northern part of the Korean Peninsula, where "gari" means "beef rib." Nowadays, brisket is commonly used instead of beef rib in this dish.

Chicken Corn Soup, Chinese Style

Total Time: 30 minutes

Ingredients (for 4 persons):
- 8 ounces of chicken breast
- 1 package of corn, Goya (1 pound)
- ½ tablespoon of Kosher salt
- ⅔ -inch piece of ginger
- ¼ stalk of big green onion stalk
- ½ tablespoon of rice wine
- 1 tablespoon of grapeseed oil
- 1 ½ quarts of coconut water
- 3 tablespoons of watery starch
- 2 egg whites

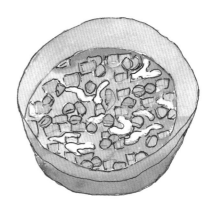

Method:

1. Begin by preparing a large pot and bringing water to a boil. Add the chicken breast along with ginger, salt, and the big green onion stalk. In the middle of the cooking process, introduce the rice wine to eliminate any unwanted chicken odors.

2. Once cooked, remove the chicken breast from the pot and rinse it with cold water to eliminate excess fat.

3. Proceed to chop and dice the chicken breast from step 2, then set it aside.

4. Heat a frying pan and add the grapeseed oil. Once the oil is hot, stir-fry the pieces of ginger and the big green onion.

5. Once the flavors of ginger and green onion have infused into the oil, remove them. Add the rice wine and coconut water to the pan and bring it to a boil.

6. As the mixture from step 5 boils, add the corn, the diced chicken breast from step 3, and the salt. Continue boiling.

7. Dilute the watery starch with an equal amount of water and add half of the mixture to the corn soup. Gradually add the remaining watery starch mixture to control the thickness of the corn soup.

8. Whisk the egg whites and gently add them, bit by bit, on top of the soup.

Clear Stew with Codfish Head (Daegu Meoritang / 대구머리탕)

Total Time: 50 minutes

Ingredients (for 4 persons):

- 2 pounds of codfish head
- 10 large dried anchovies
- 4 large-eyed herrings
- 1 sheet of kelp (about 4"x 4", palm size)
- ⅓ small radish, thickly chopped
- 3 leaves of Napa cabbage, cut into 2-inch pieces
- ½ medium-sized onion
- 1-inch piece of ginger, sliced
- 3 cloves of garlic, thinly sliced
- ¼ stalk of big green onion, chopped
- ½ medium carrot
- 1 red pepper (1 ounce)
- 1 tablespoon of coarse sea salt
- ½ teaspoon of soup soy sauce
- 3 tablespoons of mirin
- 1 teaspoon of pepper

Method:

1. In a 6-quart pot, bring 3 quarts of water to a boil. Add large dried anchovies, large-eyed herrings, and kelp. Boil for about 10 minutes, then strain the broth, discarding the large dried anchovies, large-eyed herring, and kelp. Save the broth.
2. Prepare all the ingredients as listed.
3. Bring the strained broth back to a boil. Add

codfish head, radish, Napa cabbage, onion, white part of the big green onion, ginger, and garlic. Boil over strong heat for 5 minutes. Season with coarse sea salt, soup soy sauce, mirin, and pepper. Reduce the heat to medium and skim off any fishy bubbles. Simmer for an additional 20 minutes. Your codfish head soup is ready to serve.

 Notes

- For this soup, you'll need a good amount of mirin and ginger compared to other soups.
- Codfish head is known for its flavorful broth, and a single head can create a rich and delicious soup.

Cold Gim Soup (Gim Guk / 김국)

Total Time: 20 minutes

Ingredients (for 4 persons):
- 3 sheets of gim
- 1 green pepper (1 ounce)
- 1 red pepper (1 ounce)
- 5 large dried anchovies
- handful amount of dry small shrimp (¼ ounce)
- ½ sheet of kelp (4"x 4")
- 4 cups of water
- 1 tablespoon of soy sauce
- ½ tablespoon of red pepper powder
- ½ tablespoon of rice vinegar
- ½ tablespoon of organic sugar
- 3 cloves of garlic, minced
- ½ teaspoon of roasted sesame seeds

Method:

1. Begin by boiling 4 cups of water with the large dried anchovies, dry small shrimp, and kelp. Once boiled, strain the broth and allow it to cool. Set aside.

2. Roast the 3 sheets of gim on a portable grill until they become slightly crispy. After roasting, break them into pieces inside a zip lock bag.

3. In a mixing bowl, combine the roasted gim, soy sauce, red pepper powder, rice vinegar, organic sugar, minced garlic, and roasted sesame seeds. Press down on the ingredients with a spoon.

4. Pour the cooled broth (from step 1) into the bowl with the other ingredients. Add some ice to help cool it down further.

5. Finally, garnish the cold soup with thinly sliced green pepper and red pepper.

Corn Cream Soup

Total Time: 30 minutes

Ingredients (for 4 children):
- 2 corn cobs
- ½ potato, peeled
- ½ yam, peeled
- 2 medium carrots, peeled
- ⅙ of a medium sweet onion
- 2 cups of whole milk
- 1 cup of water
- 4 tablespoons of salted butter
- ⅛ teaspoon of salt

Method:

1. Steam the corn cobs, potato, yam, carrot, and onion for 20 minutes.

2. Remove the corn kernels from the cobs and place them in a bowl.

3. Place the steamed ingredients from step 1 and the corn kernels in a blender. Pour in the milk and water. Add the butter and salt last. Press the soup button.

 Notes

- This recipe is a favorite of my 20-month-old grandson, Hansen.
- I highly recommend using a Vitamix Blender for making baby food. It's very convenient for young mothers preparing food for their little ones.

Crabmeat Soup (Gesal Soup / 게살수프)

Total Time: 30 minutes

Ingredients (for 4 persons):
- ½ cup of crab meat (left-over from steamed crabs at the last dinner)
- 1 bundle of enoki mushrooms (4 ounces)
- 2 cups of chicken broth
- 1 teaspoon of salt
- ¼ teaspoon of pepper
- 2 tablespoons of cornstarch
- 1 tablespoon of mirin
- 4 egg whites
- 2 stalks of scallions
- 1 tablespoon of roasted sesame oil

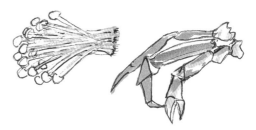

Method:

1. Begin by cutting off the root part of the enoki mushrooms and separating the stems with both hands. Cut them in half.

2. Beat the egg whites with chopsticks and set them aside.

3. Mix the chicken broth, 2 cups of water, and salt in a pot and bring it to a boil. Add the crabmeat and the enoki mushrooms from step 1 into the broth mixture along with pepper. Keep boiling for 2 minutes over medium heat.

4. Mix 1 cup of water with 2 tablespoons of cornstarch.

5. Reduce the heat to medium-low, swirl in the the liquid from #4 and the egg whites, and thoroughly mix the soup.

6. Turn off the heat, and add the roasted sesame oil, garnishing with chopped scallions. Serve your delicious crab and enoki mushroom soup.

Fishcake Soup (Eomukttang, Odeng / 어묵탕, 오뎅)

Total Time: 30 minutes

Ingredients (for 4 persons):

- 8 ounces of fish cake
- ¼ small Korean radish
- ¼ stalk of big green onion
- 2 tablespoons of soup soy sauce
- 2 tablespoons of soy sauce
- ½ teaspoon of Kosher salt
- a pinch of pepper
- a pinch of red pepper powder

Soup:

- 20 large dried anchovies
- 3x3-inch piece of kelp
- ½ onion
- 2 quarts of water

Method:

1. Parboil the fish cakes or pour boiling water over them to remove excess frying oil. Cut the fish cakes into bite-sized pieces. You can use skewers if you prefer.

2. Peel the radish, cut it into four pieces at a 1.5-inch height, and brown all sides.

3. Blanch the spinach, then make small bundles of spinach leaves, holding them tightly with their own stems.

4. Chop the big green onion.

5. Make broth by boiling water with anchovies and kelp for about 10 minutes. Remove the anchovies and kelp.

6. Boil the anchovy broth from step 5, add the fish cake pieces from step 1, reduce the heat to low, and season with soup soy sauce, salt, and pepper. Then, add the browned radish from step 2, chopped big green onion from step 4, and the bundles of spinach from step 3. Serve with red pepper powder on the table.

Hand Torn Dough Soup (Sujebi & Gamja Ongsimi / 수제비와 감자옹심이)

Total Time: 30 minutes

Ingredients (for 4 persons):
For the Dough:
- 2 cups of all-purpose flour
- ½ cup of water
- 1 teaspoon of Kosher salt

For the Soup:
- 20 large dried anchovies
- 8 large-eyed herrings
- 1 palm-sized piece of kelp (0.2 ounces)
- 1 ½ quarts of water
- 2 medium-sized potatoes, chopped
- ½ zucchini, chopped
- 1 medium carrot, chopped
- ¼ stalk of big green onion, sliced
- 1 ½ tablespoons of soup soy sauce
- ½ clove of garlic, sliced
- ⅛ teaspoon of pepper

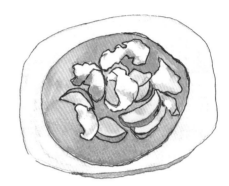

Method:

1. In a large bowl, mix the all-purpose flour and water to create a slightly watery dough. Add salt to the dough and knead it thoroughly. Refrigerate the dough for about 1 hour.
2. In a pot, bring 1.5 quarts of water to a boil. Add large dried anchovies, large-eyed herrings, and kelp. Boil for approximately 10 minutes, then remove the anchovies, herrings, and kelp.

3. Season the broth (from step 2) with soup soy sauce and add the chopped potatoes. Boil and cook the potatoes for 5 minutes.
4. To make the dough balls (Ongshimi), pull out thin pieces of dough with your three fingers on your right hand while holding the rest of the dough with your left hand. After 3 minutes, add the zucchini, carrot, garlic, big green onion, and pepper.

Tip:

You can also make Potato Ongshimi by grating potatoes, letting the potato starch settle at the bottom, and using this potato sediment to shape little balls (birds' eggs). Adding a small amount of potato starch (5%) helps to shape Ongshimi firmly and enhances their chewy texture.

Kimchi Stew (Kimchi Jjigae / 김치찌개)

Total Time: 45 minutes

Ingredients (for 4 persons):
- 2 cups of fully fermented whole kimchi, chopped
- ½ pound of pork belly
- 1 stalk of big green onion
- 1 large onion
- 2 tablespoons of mirin
- ½ teaspoon of pepper
- 1 teaspoon of roasted sesame oil
- 2 tablespoons of grapeseed oil
- about 2 cups of water
- ½ box of tofu (optional)

Method:

1. Cut the pork into pieces that are 2 inches long, 1 inch wide, and ½ inch thick. Set it aside.
2. Cut the kimchi into pieces that are 2 inches long. Set it aside.
3. Slice the onion and big green onion. Set them aside.
4. Heat a deep-frying pan and add grapeseed oil.
5. Once the grapeseed oil is heated, stir-fry the pork. Add pepper and mirin to the pork.

6. When the pork is fully cooked, add the kimchi. Keep stirring until the kimchi becomes translucent and soft. Take your time at this stage. In the middle of the process, add the sliced onion and big green onion.
7. Pour in the water and the kimchi juice from the kimchi container. Boil for about 15 minutes. Add the sesame oil last.

You can also add tofu if you like.

Mixed Bone Soup (Seolleongtang / 설렁탕)

Total Time: 6 hours

Ingredients (for 4 persons):
- 1 package of marrow bone, frozen (1 ½ pounds)
- 1 package of mixed bone, frozen (1 ½ pounds)
- ½ pounds of beef shin
- 1 pound of cow tongue, stomach, small intestine, and tripe
- ½-inch piece of ginger, diced (½ tablespoon)
- 3 cloves of garlic, diced
- 1 whole medium onion
- 1 stalk of celery, cut into ¼ inch pieces
- ¼ stalk of big green onion, cut into ¼ pieces

Green Pepper Seasoning (Dadaegi):
- 1 tablespoon of green pepper, diced
- 1 ½ cloves of garlic, diced
- 1 tablespoon of red pepper powder
- 1 tablespoon of diced shrimp sauce

Method:

1. In a large deep pot, place the bones and pour enough water to cover them. Boil for about 30 minutes. Discard this liquid.

2. Pour 6 times more water than the amount of bones over the bones and boil over high heat.

Once it starts boiling, reduce the heat to low and simmer for about 3~6 hours. Set it aside.

3. Repeat the procedure from step 2 to make the second soup. Combine the first and second soups. Boil for one more hour with a pouch containing ginger, garlic, onion, celery, and big green onion.

4. Take out the various meats, wash them with cold water, and drain. Chop the meat into biting-sized pieces. Set it aside.

5. Cook fine noodles according to package instructions, then wash them in cold water and drain. Set it aside.

6. In individual serving bowls, place the noodles and chopped meats. Pour hot soup over the noodles and meats, changing the hot soup several times. Add boiling soup last. Serve piping hot.

7. Chopped big green onion, green pepper seasoning, salt, and pepper can be provided on the dining table for seasoning according to individual preferences.

Seolleongtang is known for its rich flavor and variety of ingredients. Enjoy this comforting dish, especially on cold days!

New England Clam Chowder

Total Time: 30 minutes

Ingredients (for 4 persons):
- 1 can of clams, chopped (1 quart with juice)
- ⅓ pound of salt pork, chopped.
- 2 potatoes, diced
- 2 quarts of cold water
- 1 onion, minced
- 2 stalks of celery, minced
- ½ teaspoon of thyme
- ¼ cup of salted butter
- ¼ cup of all-purpose flour
- 2 quarts of milk
- Kosher salt (to taste)
- pepper (to taste)

Method:

1. Start by stir-frying the salt pork and potatoes, then add ¼ quart of cold water and boil until the potatoes are well cooked.

2. In a separate pan, sauté the onion, thyme, and celery with butter. Set this mixture aside.

3. In a bowl, thoroughly mix the flour and ¼ quart of cold water, ensuring there are no lumps.

4. Chop the clam meat, saving the juice separately. Do not dice the meat.

5. In a large pot, combine the sautéed ingredients from step 1 and step 2 with the chopped clams and their juice. Pour in 1.5 quarts of water, stirring as needed until the chowder reaches your desired thickness. Season with salt and pepper.

6. Add the milk and heat until it's just about to boil. Be careful not to let it boil.

You can top each serving with parsley and serve with cubed dry bread.

New York Style Clam Chowder

Total Time: 30 minutes

Ingredients (for 4 persons):
- 1 can of clams (1 quart with juice)
- ¼ pound of salt pork
- 1 medium carrot, diced
- 1 potato, diced
- 2 quarts of cold water
- 1 onion, minced
- ½ stalk of celery, minced
- 1 can of crushed tomatoes
- ½ teaspoon of thyme
- sprinkle of pepper

Method:

1. In this New York-style clam chowder, carrots are added. Start by stir-frying the salt pork, carrots, and potatoes. Add ¼ quart of cold water and boil until the potatoes are well cooked.
2. Sauté the onion and celery with the thyme until tender.
3. Chop the clam meat, saving the juice separately.

4. In a large pot, combine the sautéed ingredients from step 1 and step 2 with the chopped clams and their juice. Add 1.5 quarts of water, stirring as needed until the chowder reaches your desired thickness. Season with pepper.
5. Stir in the crushed tomatoes and heat through.

Oxtail Soup (Kkori Gomtang / 꼬리곰탕)

Total Time: 1 hour 30 minutes

Ingredients (for 3 persons):
- 1 package of oxtail (4 pounds)
- 2 quarts of water
- 2-inch piece of ginger (1 ounce)
- 3 cloves of garlic
- 2 stalks of big green onion with white roots
- 1 tablespoon of mirin

Method:

1. Begin by immersing the oxtail in cold water for about 5 hours to remove any impurities.

2. In a large pot, bring 2 quarts of water to a boil. Add the oxtail along with the ginger, big green onion, garlic, and mirin. Boil over high heat for 1 hour.

3. After the initial hour, reduce the heat to low and continue to simmer the soup for about 2 hours or until the beef surrounding the bone becomes tender and easy to pull apart. It should not be tough when you eat it.

4. To remove excess fat from the soup, add 3 to 4 cups of ice cubes. The fat will solidify and stick to the ice cubes, forming white solid chunks. Quickly remove the ice cubes with the solidified fat before they melt into the soup. Bring the soup back to a boil.

5. Serve the oxtail soup hot, garnished with chopped big green onions, and offer vinegary soy sauce on the side for dipping.

Radish Beef Soup (Mu Guk / 무국)

Total Time: 30 minutes

Ingredients (for 4 persons):
- ¼ of small Korean radish, thinly sliced
- 4 ounces of beef brisket sliced and chopped
- 3 pieces of kelp, each about 2-inch x 2-inch
- 1 leek stalk
- ½ onion
- 2 cloves of garlic, sliced
- ⅓ -inch piece ginger, sliced
- 1 tablespoon of sesame oil
- 2 teaspoons of soup soy sauce
- ½ teaspoon of Kosher salt
- 1 ½ pints of water

Method:

1. Start by slicing and chopping the beef brisket. Also, cut the radish into thin slices, about 1-inch x 1-inch in size. Prepare the kelp pieces.

2. In a pot, heat up the sesame oil over strong heat. Stir-fry the beef brisket until it's cooked.

3. Add the sliced radish to the pot and continue stir-frying. Season with the soup soy sauce for flavor.

4. Pour in the water and add the kelp pieces. Allow the mixture to come to a boil.

5. Once the water is boiling, add the sliced ginger, garlic, leek, and onion to the pot.

6. Reduce the heat to medium and let the soup simmer for about 15 minutes.

7. Remove the kelp, onion, and leek from the soup.

8. Season the soup with Kosher salt and pepper to taste.

9. Serve the radish and beef brisket soup hot, garnished with chopped scallions for extra flavor and freshness.

Rich Soybean Paste Stew (Cheonggukjang Jjigae / 청국장찌개)

Total Time: 20 minutes

Ingredients (for 2 persons):

- ½ box of tofu, cut into ½-inch cubes (5 ounces)
- 2 ounces of diced beef
- ¼ cup of kimchi, cut into 2-inch squares (2 ounces)
- 3 tablespoons of rich soybean paste (chunggukjang)
- ½ stalk of big green onion (4 ounces)
- 1 green pepper (1 ounce)
- 1 red pepper (1 ounce)
- ½ teaspoon of Kosher salt
- 1 clove of garlic, diced
- pepper (to taste)
- 4 cups of water

Method:

1. Season the diced beef with big green onion, garlic, salt, and pepper. Set it aside.
2. Stir-fry the marinated beef mixture and add 3 cups of water over low heat. Remove any bubbles to create a clear broth.
3. Take out most of the filling from the kimchi and cut it into 2-inch squares. Tearing the kimchi lengthwise with your hands can also be visually appealing.
4. Mix the rich soybean paste (chunggukjang) into 1 cup of water until well combined, and then pour it into the clear broth created in step 2. Once everything is thoroughly mixed, add the kimchi, tofu, big green onion, green pepper, and red pepper. Boil until the soup turns into a stew-like consistency and season with salt. The stew should be thick, not diluted.

Make sure to find a high-quality rich soybean paste (chunggukjang) with a gentle aroma and rich flavor for the best result.

Salted Pollock Roe Stew (Myeongran Jjigae / 명란찌개)

Total Time: 20 minutes

Ingredients (for 2 persons):
- 2 salted pollock roe, cut into 1-inch lengths (2.5 ounces total)
- ½ box of tofu, cut into ½-inch cubes (5 ounces)
- ½ zucchini, chopped
- ¼ sweet onion, chopped
- 2 stalks of scallion, sliced
- 1 ½ cloves of garlic, minced
- ½ tablespoon of shrimp sauce
- 2 cups of anchovy broth

Method:

1. Start by boiling the anchovy broth. Add zucchini, onion, scallion, and garlic to the boiling broth.
2. Add the 2-inch pieces of salted pollock roe and tofu. Season the stew with a little shrimp sauce.

Tip:

If you have kept salted pollock roe in your refrigerator for over a week, and it has lost its fresh taste, don't worry. It can still be used to make a flavorful pollock roe stew.

Sesame Dried Pollock Soup (Bugeotguk / 북엇국)

Total Time: 20 minutes

Ingredients (for 2 persons):

- 1 cup of dried pollock strips
- 2 teaspoons of roasted sesame oil
- 3 cups of water
- 1 egg
- ⅓ stalk of big green onion (⅓ cup)
- 2 cloves of garlic, sliced
- ⅓-inch piece of ginger, sliced
- dash of pepper
- 1 tablespoon of shrimp sauce

Method:

1. In a pot, begin by stir-frying the dried pollock strips with the roasted sesame oil.
2. Add the 3 cups of water to the pot and bring it to a boil. Allow it to simmer.
3. While the soup is simmering, beat an egg and mix it with the chopped big green onion.
4. Pour the egg and green onion mixture into the simmering soup (#1).
5. Serve your sesame dried pollock soup hot.
6. At the dining table, season individual servings with shrimp sauce to taste.

Seaweed Soup (Miyeok Guk / 미역국)

Total Time: 30 minutes

Ingredients (for 4 persons):

- ⅔ ounce of dried seaweed
- 6 ounces of thick-sliced beef, cut into 1-inch pieces
- 7 cups of water
- 1 clove of garlic, diced
- 2 tablespoons of roasted sesame seed oil
- 1 ½ tablespoons of soup soy sauce
- ½ teaspoon of Kosher salt

Method:

1. Soak the dried seaweed in water for over 10 minutes until it swells to about 8 times its original volume. Scrub the seaweed thoroughly until no bubbles come out of it. Wash it and place the cleaned seaweed on a cutting board. Cut it into 2-inch-wide strips.

2. Heat a 4-quart pot, add the sesame seed oil, and stir-fry the beef slices until cooked. Then, add the cut seaweed and garlic. Stir well to prevent anything from sticking to the bottom of the pot. Continue stirring until the seaweed becomes very soft and loses its stiffness.

3. Pour the water into the pot and add the soup soy sauce and salt. Start at high heat, but once it boils, reduce the heat to medium and let it simmer for 30 minutes with the lid on. This process allows the flavors to meld, resulting in a rich and flavorful seaweed soup. Serve hot.

Tip:

If you prefer a clear seaweed soup, you can skip stir-frying the beef. Instead, simply add a piece of shin meat to the boiling water until it's fully cooked, then add the soaked seaweed. Simmer for over an hour on medium-low heat with the lid on. Remove the meat, and you can use it as a garnish for the soup, either tearing it apart or slicing it thinly.

Soft Tofu Stew (Sundubu Jjigae / 순두부찌개)

Total Time: 15 minutes

Ingredients (for 2 persons):

- 1 package of soft tofu (11 ounces)
- 4 ounces of chopped pork
- ⅔ -inch piece of ginger, sliced
- ¼ onion, chopped
- ⅛ stalk of big green onion, chopped
- 4 ounces of raw shiitake mushrooms, sliced
- ¼ red pepper, chopped (¼ ounce)
- 2 cloves of garlic, diced
- 1 tablespoon of roasted sesame oil
- 1 teaspoon of red pepper powder
- 1 ½ teaspoons of shrimp sauce
- 1 ½ cups of anchovy broth
- 1 egg

Method:

1. Open the package of soft tofu and cut it in half. Set it aside.

2. Heat a frying pan over high heat, then add sesame oil, big green onion, and garlic. Stir-fry until fragrant.

3. Add the chopped pork and ginger to the pan. Continue to stir-fry until the pork releases its oil and begins to cook.

4. Reduce the heat to low and add the red pepper and red pepper powder. Stir everything together.

5. Pour in the anchovy broth, then add the shrimp sauce, raw shiitake mushrooms, and chopped onion. Let the mixture come to a boil.

6. Once the vegetables in the stew are boiling, gently scoop out several pieces of soft tofu with a flat ladle and place them on the surface of the stew like floating boats. Do not stir the tofu into the stew; simply let it simmer on top.

7. Break an egg and separate the yolk from the white. Whisk the egg white and spread it evenly around the pot. Place the egg yolk in the center.

8. The soft tofu stew is now ready to eat!

It's recommended to use a small earthen pot for cooking the best soft tofu stew at home. You can find different sizes of earthen pots, with medium size suitable for serving 4 persons and small size for serving 2 persons. Additionally, if you have leftover anchovy broth (boiled and filtered from anchovies, large-eyed herrings, or kelp), you can store it in the refrigerator for future use. It's convenient for making flavorful broths like this one.

Soybean Paste Soup with Spinach & Soybean Sprout
(Doenjangguk / 된장국)

Total Time: 20 minutes

Ingredients (for 2 persons):

- 1 ½ tablespoons of soybean paste
- sprinkle of Kosher salt
- 2 cups of rice water
- 6 large dried anchovies
- 2 large-eyed herrings
- ¼ palm-sized piece of kelp
- 1 clove of garlic
- ⅓ bunch of spinach blanched and washed (½ cup)
- 1 ounce of soybean sprouts cleaned and washed
- ¼ of onion
- 1 stalk of scallion
- ⅕ of red pepper (½ ounce)
- ⅛ box of tofu (2 ounces)

Method:

1. When the rice water boils, add large dried anchovies, large-eyed herrings, and kelp to create the soup base. Boil for 7-10 minutes, then remove the anchovies, pilchards, and kelp.
2. Blanch the spinach, rinse it with cold running water, and drain.
3. In the soup base from step 1, dissolve the soybean paste first. Then, add the blanched spinach, soybean sprouts, and all the remaining vegetables. Add the tofu and red pepper last.

 Notes

- If you prefer, you can make your own version of this soup by using each vegetable individually or combining them with ingredients like mushrooms, radishes, dried cabbage leaves, and dried radish leaves.
- While homemade soybean paste is the best choice, it may not be readily available. In that case, look for a commercial product that closely matches the taste, ideally not too salty or sweet.

Soybean Paste Stew (Doenjang Jjigae / 된장찌개)

Total Time: 20 minutes

Ingredients (for 2 persons):
- 1 ½ tablespoons of soybean paste
- 1 ½ cups of rice water
- 6 large dried anchovies
- 2 large-eyed herrings
- kelp, approximately ¼ palm size
- 2 cloves of garlic
- 1 ½ ounces of radish
- ½ zucchini
- ¼ onion
- 1 fresh shiitake mushroom
- 1 stalk of scallion
- ½ red pepper (½ ounce)
- ⅕ box of tofu (2 ounces)

Method:

1. Begin by bringing the rice water to a boil. Once it boils, add the large dried anchovies and pilchards, and create a soup base by boiling for 5-7 minutes with the kelp. Remove the anchovies, large-eyed herrings, and kelp from the soup.

2. Prepare the vegetables by cutting the radish into thin, 1-inch square slices. Chop the zucchini and onion.

3. Cut the fresh shiitake mushroom in half, slice the scallions into 1.5-inch lengths, and cube the tofu into 1-inch pieces.

4. In the soup base from step 1, dissolve the soybean paste first. Then, add the prepared vegetables (radish, zucchini, onion), along with the fresh shiitake mushrooms and tofu. Sprinkle the red pepper on top.

5. Boil the stew over medium heat for about 5 minutes.

6. Serve the Korean soybean paste stew while it's still boiling on the dining table for the most authentic experience.

 Notes

Homemade soybean paste is the best choice, but it can be challenging to find families that make their own soybean paste at home these days. Look for a commercial product that comes closest in taste to authentic Korean soybean paste.

Spicy Beef & Vegetable Soup (Yukgaejang / 육개장)

Total Time: 1 hour 30 minutes

Ingredients (for 4 persons):

- 1.5 pounds of beef flank steak
- 3 quarts of water
- 2 bundles of big green onion (4 pounds)
- 1 ounce of dry bracken
- ½ ounce of dry taro stem
- half of a big onion
- 1-inch of ginger, sliced
- 9 cloves of minced garlic
- 3 tablespoons of roasted sesame oil
- 3 tablespoons of red pepper powder
- 2 tablespoons of soup soy sauce
- ½ teaspoon of Kosher salt
- 1 tablespoon of mirin
- 1 teaspoon of roasted sesame seeds, crushed
- ⅛ teaspoon of pepper

Method:

1. Start by parboiling the dry bracken and taro stem separately. Then, immerse them in cold water overnight, changing the water several times. The next day, drain them and set them aside.

2. In a pot, bring water to a boil and add ginger, mirin, and beef. Boil over high heat for about 1 to 1.5 hours. Filter the broth using a paper towel-lined sieve on a basket. This step is crucial to prevent the broth from becoming cloudy and maintaining a clean taste.

3. Cut the beef into 2 or 3 pieces, about the size of your pinky finger. Be careful not to make them too thin. Each piece should be about 3 inches long.

4. Remove the fat from the broth. You can do this by letting the broth cool in a cold place or using plenty of ice cubes to solidify the fat, which can then be easily removed.

5. Cut all the big green onions into 3-inch

lengths. Blanch them, wash them with cold running water, and drain. Set them aside.

6. In a large bowl, combine the soaked bracken and taro stem (step 1), sliced beef (step 3), and de-fatted broth (step 4). Add pepper, sesame seeds, salt, and soup soy sauce.

7. Prepare the red pepper oil: Heat sesame oil in a frying pan. Once the oil is hot, add garlic, turn off the heat, and add red pepper powder. Mix well. Stir-frying the red pepper powder with the flame on can result in a dull-colored broth with less flavor.

8. Pour the hot red pepper oil (#7) onto the blanched green onions (#5) and the mixture in the bowl (#6). Mix everything together with your hands. Bring the broth to a boil again, and then add the mixed contents. Boil until all the ingredients float up.

Vegetable Soup (Yachae Soup / 야채수프)

Total Time: 30 minutes

Ingredients (for 3 persons):

- 4 ounces of beef, cut into chunks
- ⅛ of a cabbage, cut into chunks
- 2 medium-sized potatoes, cut into chunks
- ½ of a medium-sized onion, chopped
- 1 medium-sized carrot, chopped
- 1 stalk of celery, chopped
- 2 cloves of garlic, sliced
- 1 large ripe tomato, chopped
- 2 tablespoons of tomato paste
- ¼ teaspoon of Kosher salt
- 1 tablespoon of unsalted butter
- 2 quarts of water
- ⅛ teaspoon of pepper
- parsley for garnish

Method:

1. Begin by preparing all the vegetables as specified in the ingredient list. Set them aside.
2. Heat a saucepan over high heat. Once the pot is hot, melt the butter. Start by stir-frying the garlic and onion until they become fragrant. Then, add the beef and continue stir-frying. Place all the vegetables listed in step 1, along with the tomato paste, on top of the beef, garlic, and onion. Keep stir-frying until they are all cooked.
3. Pour in the water and allow it to come to a boil. Reduce the heat to medium, cover with a lid, and simmer for approximately 2 hours. During this time, the water will reduce to about ⅘ of its original volume through evaporation. Season the stew with salt.
4. Serve the delicious stew garnished with parsley.

Tip:

Instead of using butter, you have the option to use olive oil. However, your preference for butter is perfectly fine and adds a unique flavor to the dish.

Wild Pollock Stew (Maeuntang / 매운탕)

Total Time: 30 minutes

Ingredients:

- 1.2 pounds wild pollock, cut into 2-inch pieces
- 5 large dried anchovies
- 2 large-eyed herrings
- 3x1 inch piece of kelp (approximately half palm size)
- ⅓ medium-sized radish, sliced into ½-inch pieces
- 1 tablespoon red pepper powder
- 2 Napa cabbage leaves, sliced into 2-inch pieces
- ¼ stalk of big green onion, cut into 2-inch pieces
- ¼ onion, sliced
- 3 cloves of garlic, thinly sliced
- ½ teaspoon ginger, sliced
- 1 green pepper sliced sideways (1 ounce)
- 1 red pepper sliced sideways (1 ounce)
- 1 tablespoon Kosher salt
- ½ teaspoon soup soy sauce
- 1 teaspoon cooking wine or mirin
- ⅕ bunch of watercress (2 ounces)
- ⅕ box of tofu (2 ounces)
- ⅕ bunch of crown daisy (2 ounces)

Method:

1. In a pot, boil 1.5 quarts of water with large dried anchovies, large-eyed herring, and kelp for 10 minutes to create anchovy broth. Remove and discard these three ingredients. Set the broth aside.

2. Sprinkle sea salt on the pieces of pollock to firm up the surface and help retain the fish's juices. Rinse the salt off under cold running water and drain. Set aside.

3. Arrange all the vegetables as listed in the ingredients on a tray.

4. Bring the anchovy broth back to a boil. Mix in and dissolve the red pepper powder and salt. Add the pollock, radish, big green onion, garlic, ginger, red pepper, watercress, and green pepper. Pour in the mirin and boil over high heat for about 5 minutes. Skim off any scum that forms on the surface.

5. Reduce the heat to medium and continue boiling for another 10 minutes.

6. Finally, add the tofu and soup soy sauce. Serve garnished with crown daisy.

Chapter V

Kimchi & Pickles

Boiled Pickled Cucumber (Oi Sukjangajji / 오이숙장아찌)

Total Time: 2 days for pickling

Ingredients:

- 1 pound of cucumbers
  (pickling cucumbers or English cucumbers)
- 2 tablespoons of Kosher salt
- ¾ cup of soy sauce
- ⅜ cup of organic sugar
- ¼ cup of white vinegar
- ¾ cup of water
- ¼ cup of soju

Method:

1. Scrub the cucumbers, rub them with salt, and then rinse them. Place the cucumbers in a container with a lid.
2. Prepare the sauce using the quantities listed above.
3. Boil the sauce and pour it over the cucumbers while it's still boiling. This helps keep the pickled cucumbers crispy. Place wooden chopsticks on top of the cucumbers and weigh them down with something heavy, like a flat stone. Let them stand overnight. The next day, remove the cucumbers and save the sauce. Place the cucumbers back in the container, changing the positions of the upper ones to the bottom.

4. Boil the saved sauce, then allow it to cool completely before pouring it over the cucumbers again. Keep the container in the refrigerator.

 Notes

When serving pickled cucumbers, cut them into bite-sized pieces and sprinkle with sesame seeds, or mix them with sesame oil, red pepper powder, and chopped big green onions.
Soju is used to prevent the growth of fungi in this recipe.

Cabbage Kimchi (Yangbaechu Kimchi / 양배추김치)

Total Time: 1 hour and 30 minutes

Ingredients:
- ½ cabbage (approximately 2 pounds)
- ½ cup of coarse sea salt
- 2 quarts of water
- ¼ stalk of chopped big green onions
- ¼ of sweet onion
- 3 cloves of garlic
- ⅓ -inch piece of ginger
- ¼ cup of red pepper flakes
- ½ apple
- 2 teaspoons of shrimp sauce
- 1 teaspoon of anchovy sauce
- ¼ cup of flour

Method:

1. Start by removing the root portion of the cabbage and then cut the cabbage leaves into 2-inch squares.

2. Sprinkle half of the coarse sea salt onto the cabbage leaves. In a separate container, dissolve the remaining salt in water and pour this saltwater over the cabbage leaves. Wait for approximately 30 minutes, then turn the cabbage leaves upside down. After about 10 minutes, rinse the cabbage leaves thoroughly under cold running water. Drain them and set them aside.

3. Use a small garlic chopper to finely grind the sweet onion, garlic, ginger, and apple. Also, chop the big green onion.

4. In a large bowl, combine the mixture from step 3 with the shrimp sauce, anchovy sauce, and flour porridge. Mix this well with the cabbage from step 2. Allow the mixture to ferment at room temperature for one or two days.

Chayote Pickle (Chayote Jangajji / 차요테장아찌)

Total Time: 4~6 days for pickling

Ingredients:

- 4 chayotes (2 pounds)
- ½ cup of soy sauce
- 1 cup of water
- ½ cup of white vinegar
- ½ cup of organic sugar
- ½ red pepper (½ ounce)

Method:

1. Cut the chayote into four pieces lengthwise. Remove the seeded part and cut each piece into a triangular fan shape, about ⅛ -inch thick. Place them in a large-mouthed bottle with a lid.
2. Mix soy sauce, water, vinegar, and sugar. You can adjust the ratio of vinegar and sugar according to your taste.
3. In a deep pot, boil soy sauce, water, and sugar. Once it boils, add the vinegar, and continue boiling. Allow it to cool, then pour it over the chayote pieces. To prevent the chayote pieces from floating to the top, cover the surface with something heavy. Store it in the refrigerator and let it sit for a day or two.

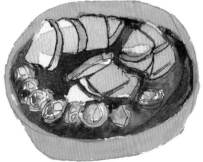

4. About 3-4 days later, you can boil the liquid from step 3 again, let it cool, and pour it over the chayote pieces for extra flavor.
5. You can also add radish, onion, or jalapeno to complement the chayote.

Chives Kimchi (Buchu Kimchi / 부추김치)

Total Time: 30 minutes

Ingredients:
- ½ bundle of chives (10 ounces)
- ¾ cup of red pepper flakes
- ¼ cup of anchovy extract
- 1 tablespoon of organic sugar

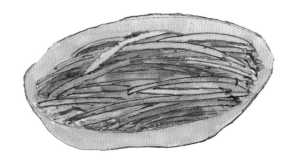

Method:

1. Select a bundle of fresh chives, preferably without wilted leaves. Trim each chive neatly, removing 1 inch of the white ends. Wash and drain the chives, then set them aside.

2. Mix anchovy extract, red pepper flakes, and sugar thoroughly.

3. In a wide bowl, coat and mix a handful of chive leaves with some of the mixture from step 2. Transfer them to one side and repeat this process until all chives have been coated and mixed. Combine them all and place them in a container. Leave at room temperature for one or two days.

Tip:

Do not cut the chives leaves; simply trim the white ends when making chives kimchi.

When preparing chives kimchi, you don't need garlic, ginger, or big green onion. Chives alone provide these flavors. You only require a small amount of sugar for the best taste.

Cucumber Pickle (Oiji / 오이지)

Total Time: 10 days for pickling

I. Ingredients: Traditional Way
- 10 thin & long cucumbers (4 pounds)
- 1 ¼ cups of coarse sea salt
- ¼ cup of soju
- 2 tablespoons of red pepper seeds
- 1 quart of water

Method:

1. Immerse the cucumbers in cold water for about 10 minutes to remove any possible pesticides. Then, wash them in cold running water and drain.

2. Add the salt to the water and boil. Parboil each cucumber in this saltwater for 30 seconds. Drain and set them aside.

3. In a container with a lid, place the cucumbers side by side. Pour the cooled saltwater from step 2 over the cucumbers, add soju, and add the pouch of red pepper seeds. After ten days, the cucumbers should turn into a yellowish pickle. If there's still some green tint, it means they were not salty enough, and it's considered a failure. This pickle is very salty, so when you use it, take out the excess salt by immersing it in water for about 10 minutes.

Total Time: 1 week

II. Ingredients: Contemporary Way
- 10 thin & long cucumbers (4 pound)
- ½ cup of coarse sea salt
- 1 cup of organic sugar
- 1 cup of white vinegar
- 2 ½ ounces of soju
- 2 green peppers (cut in half)

Method:

1. Immerse the cucumbers in cold water for about 10 minutes to remove any possible pesticides. Then, wash them in cold running water and drain. Set them aside.

2. In a 2-gallon zip lock bag, place the cucumbers side by side. Add the coarse sea salt, sugar, vinegar, green pepper, and soju. Seal the zip lock bag and keep the cucumber mixture at room temperature. Every other day, flip it over.

3. After one week, it will be done. Keep it in the refrigerator. The ratio of salt, sugar, and vinegar (1:2:2) helps develop the taste of the cucumber pickle.

 Note

The soju works as a preservative; the alcohol will evaporate.

Fresh Baby Napa Kimchi (Eolgari Geotjeori / 얼가리겉절이)

Total Time: 2 hours

Ingredients:

- 2 bundles of baby Napa cabbage (1 ½ pounds)
- ¾ cup of coarse sea salt
- 3 quarts of water
- ¼ stalk of big green onion
- ½ medium onion
- 2 cloves of garlic
- ⅓ -inch piece of ginger
- ¼ cup of red pepper flakes
- 1 green pepper (1 ounce)
- ¼ pear
- 2 tablespoons of anchovy sauce
- 2 tablespoons of plum extract
- ½ tablespoon of roughly ground Sichuan pepper (optional)
- 1 tablespoon of roasted sesame seeds

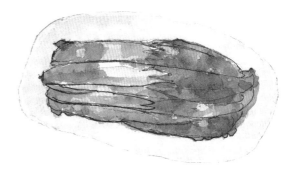

Method:

1. Start by preparing to brine the baby Napa cabbage. Sprinkle ¾ cup of coarse sea salt or more evenly over the cabbage. Pay more attention to the root part while using less salt on the leafy part. Dissolve any remaining salt in warm water and use it to immerse the whole cabbage. Allow them to be brined for 45 minutes, flipping them over once after the first 30 minutes.

2. After being brined, rinse the cabbage thoroughly under fast-running water. Hold the root part with your hand and shake it vigorously to wash away the salt. There's no need to wash the cabbages twice.

3. Drain the washed baby cabbage in a big basket. Remove the root part using a knife, but do not cut any other parts, long or short.

4. In a separate bowl, chop the garlic, ginger, onion, and pear into small pieces using a garlic chopper. Slice the big green onion and green pepper.

5. In a big bowl, combine the washed cabbage with the chopped garlic, ginger, onion, pear, red pepper flakes, anchovy sauce, and (optionally) Sichuan pepper. Sprinkle roasted sesame seeds on top.

6. Mix all the ingredients thoroughly, making sure the seasoning is evenly distributed.

7. Transfer the prepared mixture into a clean, airtight container or jars, pressing it down to eliminate air gaps.

8. Seal the container or jars and let the kimchi ferment at room temperature for a period ranging from a few days to a few weeks, depending on your desired level of fermentation. Check it regularly to taste and see if it has reached the desired flavor. Once it's fermented to your liking, you can refrigerate it to slow down the fermentation process.

Enjoy baby Napa cabbage kimchi! It's a flavorful and spicy condiment that can be served with a variety of dishes or enjoyed on its own.

Fresh Baechu Kimchi (Baechu Geotjeori / 배추겉절이)

Total Time: 2 hours

Ingredients:
- 1 Napa cabbage
- ½ stalk of big green onion
- 3 ounces of red pepper flake
- 1 small red pepper (3 ounces)
- ⅓ Fuji apple
- ⅓ Asian pear
- 2 cloves of garlic
- ⅓-inch piece of ginger
- 2 tablespoons of shrimp sauce
- 1 cup of coarse sea salt
- 1 teaspoon of roasted sesame seed
- ⅓ bundle of chives or water parsley
- 1 teaspoon of pine nuts

Method:

1. Cut the Napa cabbage to a depth of 2 inches from the root and tear the whole cabbage in half with your hands. Make another 2-inch cut into the root part of each half of the cabbage.

2. Dip the Napa cabbage halves into warm water, then remove them, and sprinkle some of the sea salt near the root part and on the leafy parts, with less salt going on the leafy parts. Let them rest in a wide container. Dissolve the remaining salt in 4 quarts of water and pour it over the Napa cabbage halves.

3. After one hour, flip the cabbage halves over and wait for an additional 15~20 minutes.

4. Holding the root part of each cabbage half under fast-running water, shake the cabbage vigorously just once. Drain them into a wide basket and set it aside.

5. Use a food processor to grind the garlic, ginger, onion, red pepper, apple, and pear. Slice the white part of the big green onion and chop the green leafy part. Set them aside.

6. Remove the root part of the cabbage and tear

the cabbage leaves lengthwise, then cut them crosswise from time to time. In a large bowl, mix these well: step 5 contents, red pepper flakes, chives, and shrimp sauce. Sprinkle sesame seeds and pine nuts.

7. You can add sesame oil if desired. I choose not to add sesame oil as it can diminish the fresh taste of "Fresh Kimchi."

Fresh Chives Kimchi (Buchu Geotjeori / 부추겉절이)

Total Time: 30 minutes

Ingredients:
- ¼ bundle of chives
- ¼ medium sweet onion
- ¼ cucumber
- 3 tablespoons of red pepper powder
- 1 tablespoon of organic sugar
- 1 tablespoon of anchovy extract
- 1 teaspoon of roasted sesame seeds, crushed
- 2 teaspoons of roasted sesame oil

Method:

1. Cut 1-inch of the white part of the chives, trim the whole leaves, wash, and drain. Set aside.
2. Slice the onion and cut the cucumber in half lengthwise, then slice it sideways.
3. In a medium bowl, mix the red pepper powder, sugar, anchovy extract, sesame seeds, and sesame oil.
4. Cut the chives into 2-inch lengths.
5. Add the chives, sliced onion, and cucumber to the mixture in step 3. Mix everything well. Serve.

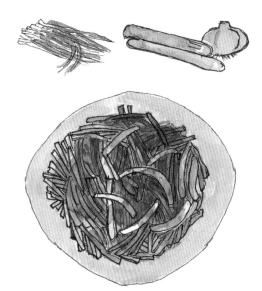

Fresh Tomato Kimchi (Tomato Geotjeori / 토마토겉절이)

Total Time: 30 minutes

Ingredients:

- 4 tomatoes
- 1 medium fuji apple
- half of a sweet onion
- 2 cloves of garlic
- 1 teaspoon of ginger juice
- ⅓ bundle of chives
- 2 tablespoons of shrimp sauce
- 1 teaspoon of organic sugar
- 1 tablespoon of roasted soybean powder
- 2 ½ tablespoons of red pepper flakes
- 1 tablespoon of coarse sea salt

Method:

1. Begin by selecting firm stem tomatoes; they should not be too big or too small.

2. Chop and grind ⅓ of the apple, onion, garlic, and salted shrimp. Set this mixture aside.

3. Cut the remaining ⅔ of the apple into ½-inch-sized cubes. Salt them with coarse salt for 2 minutes, then wash and drain them in a basket. Set them aside.

4. Cut the tomatoes in half and then into 6 pieces each. Set them aside.

5. Wash the chives under running water, remove the white bottom part, and cut them into 2-inch lengths. Set them aside.

6. In a large bowl, combine the tomatoes from step 4, the cubed apples from step 3, and the mixture from step 2. Add ginger juice, sugar, roasted soybean powder, and red pepper flakes. Gently toss all the ingredients together.

7. Add the chives to the mixture last. Adding the chives at the last moment is a tip to make delicious tomato kimchi.

Pickled Green Pepper (Putgochu Jangajji / 풋고추장아찌)

Total Time: 10 days for pickling

Ingredients:

- 12 green peppers (12 ounces total)
- 9 tablespoons of soy sauce
- ¼ cup of organic sugar
- 3 tablespoons of white vinegar
- 3 tablespoons of soju
- 9 tablespoons of water
- ⅓ ounce of kelp

Method:

1. Cut out the stem of the green peppers, leaving about ½-inch from the top. Wash the peppers under cold running water, drain them, and pat them dry with paper towels or a cotton cloth. Also, trim the tips of the peppers so that the sauce can penetrate them.

2. Place the green peppers in a container with a lid.

3. Prepare the sauce using the quantities listed above. Boil the sauce for 5 minutes with the kelp. Remove and discard the kelp. After the sauce has cooled to room temperature, pour it over the green peppers. Stir the green peppers, making sure to move the bottom ones to the top. Place something heavy on the top to keep the peppers submerged, then close the lid.

4. Let the green peppers sit at room temperature for 3 days; they will turn yellow after 7 days. After this initial period, store them in the refrigerator.

 Note

Soju, a Korean liquor with 16.9% alcohol content, is used to prevent the growth of fungi in this recipe. Pickled Green Pepper (Putgochu-jangajji).

Pre-cut Kimchi (Mak Kimchi / 막김치)

Total Time: 1 hour

Ingredients:

- 1 medium-sized Napa cabbage (2 pounds)
- 1 cup of coarse sea salt
- 4 quarts of water
- 1 clove of garlic
- ⅓-inch piece of ginger
- ½ medium-sized onion
- ¼ stalk of big green onion
- ⅓ of a big apple
- ¼ of a large-sized pear
- ⅓ cup of red pepper flakes
- 2 tablespoons of fish sauce (optional)
- 1 tablespoon of shrimp sauce (optional)
- 6 medium-sized fresh shrimp (optional)

Method:

1. Begin by cutting the root part of the cabbage lengthwise, about 2-3 inches deep, and open it into 2 pieces with your hands. Make a second smaller cut in the root part; this is simply for salting.

2. For the brine, use more salt on the root part and less salt on the upper leafy part. The more salt you use, the less time is required for salting. To be properly brined (cabbage leaves should be limp), 2-3 hours are enough if you use enough salt. Don't worry, all the salt will be washed away after it hardens the surface of cabbage leaves. Coarse sea salt does not penetrate the inner part of cabbage leaves unless left for too many hours.

3. Chop the garlic, ginger, onion, apple, and pear together in a food processor. Cut the big green onion into 1.5-inch lengths. Chop the shrimp.

4. Wash the brined cabbage under very fast-flowing water, holding the root part with your hand and shaking it vigorously. This way, you wash the cabbage only once. They won't be salty at all.

5. Drain the cabbage in a big flat basket. After draining, cut the cabbage leaves into 2-inch squares.

6. In a large bowl, mix the cabbage squares with red pepper flakes, fish sauce, shrimp sauce, garlic, ginger, big green onion, and shrimp. You can make delicious Kimchi without using fish sauce, shrimp sauce, or shrimp.

7. Fill a large container with this #6 kimchi mixture and cover it with an airtight lid. Leave some empty space at the top because the kimchi will expand during fermentation.

8. Leave your new kimchi at room temperature for about one day. This is how you allow the kimchi to start fermenting. When bubbles start rising to the surface, you can refrigerate it to keep your desired taste.

Radish Kimchi (Kkakdugi / 깍두기)

Total Time: 1 hour

Ingredients:

- 2 large Jeju Island winter radishes (6 pounds)
- ½ cup of coarse sea salt
- 2 tablespoons of red pepper flakes
- 1 package of red pepper (5 ounces)
- 1 stalk of big green onion
- 6 cloves of garlic
- 1-inch piece of ginger
- 1 ½ large pears
- 1 ½ large apples
- 2 tablespoons of shrimp sauce
- ½ cup of rice paste
  (2 tablespoons rice powder + ½ cup water)

Method:

1. Wash and clean the radishes and peel the outer layer. Place one radish on a cutting board lengthwise, divide it into several pieces with a height of ¾ inch, cut each piece into ¾-inch length and ¾-inch width to make ¾-inch cubes. Repeat the same process with the second radish. These cubes look attractive and are easy to eat.

2. Dissolve rice powder into water to make rice paste by stirring and boil with water.

Alternatively, you can use cooked rice instead. Cool the rice paste and set it aside.

3. In a big bowl, sprinkle coarse sea salt evenly over the radish cubes. Wait for 30 minutes, then wash them with a large amount of fast-running cold water just once. If the washing container has a large hole on the upper part, it's helpful to wash out the coarse sea salt from the surface of the radish cubes. Make sure not to salt the radish for more than 30 minutes.

4. Chop and grind pear, apple, red pepper flake, red pepper, garlic, ginger, shrimp sauce, and rice paste in a large food processor. Set it aside.

5. Slice the white part of the big green onion thinly and chop the green leafy part.
6. In a large and wide bowl, mix the washed radish cubes from step 3, the mixture from step 4, and the green onions from step 5. Your kkakdugi, or radish kimchi, is now ready. I don't use any sugar but use some fruits to add sweetness to the kkakdugi. Big amount!

Spicy Stuffed Cucumber Kimchi (Oi Sobagi / 오이소박이)

Total Time: 1 hour 30 minutes

Ingredients:

- 5 cucumbers, long & thin shaped (2 pounds)
- 1 tablespoon of coarse sea salt (3% saltiness)
- 1 pint of water
- ⅙ bundle of chives
- half of a pear, sliced
- ¼ sweet onion, sliced
- ⅜ cup of red pepper flakes
- 3 cloves of garlic, minced
- ½-inch piece of ginger, minced
- ¼ stalk of big green onion, chopped
- 1 tablespoon of shrimp sauce

Method:

1. Prepare a 9"x12"x1.5" cuboid container. Fill it with 1 pint of water and 1 tablespoon of coarse sea salt. Immerse the cucumbers in a single layer and leave them for about 2 hours. Wash the cucumbers in fast-running water, drain, and set them aside.

2. Once the cucumbers are fully drained, cut out both ends and slice them very thinly. Cut each cucumber into quarters. Then, cut each cucumber piece vertically into quarters, leaving ¼-inch of one end. Press and remove water from the cucumbers using several pieces of paper towels. This tip ensures crunchy cucumber kimchi. Avoid parboiling the cucumbers in salt water, as it breaks down the enzymes needed for fermentation.

3. Cut the chives into 1.5-inch lengths, slice the pear and onion, and prepare the garlic, ginger, and big green onion.

4. Mix the #3 mixture (garlic, ginger, big green onion) with the cucumber end-slices and shrimp sauce. Pack this mixture into the openings of the cucumbers. Place the stuffed cucumbers in the container and leave the remaining stuffing at the

bottom of the container alongside the stuffed cucumber pieces.

5. About one day later, it will start fermenting. Keep it in the refrigerator and enjoy it. It will remain crunchy.

 Note

The saltiness of seawater is typically around 3.1% to 3.5%. Several centuries ago, our Korean ancestors understood the principle of brining and practiced brining to harden the surfaces of cabbages, small radishes, and cucumbers with salt in seawater. They did not cut these vegetables. After the brine was done, they simply washed away all the salty seawater. They knew how to make kimchi in a way that's not salty, and we respect the wisdom of our ancestors.

Sauerkraut

Total Time: 2 weeks for fermenting

Ingredients:
- 1 medium-sized cabbage (3 ½ pounds)
- 3 tablespoons coarse sea salt

Method:

1. Begin by peeling off the outer leaves of the cabbage and thoroughly cleaning the entire cabbage.

2. Chop the cabbage into manageable pieces or shred it into strips.

3. Evenly sprinkle the coarse sea salt over the chopped or shredded cabbage.

4. Now, take a potato masher or a Cabbage Crusher and vigorously beat the salted cabbage for about 10 minutes. Apply strong pressure while doing this. As you start beating the cabbage, you'll notice a significant amount of juice being released from the cabbage itself. Stop beating when there's enough juice to cover the cabbage inside the jar. Pack and press down the shredded cabbage tightly into the jar. If the juice isn't sufficient to cover the top of the cabbage, you can add a bit of water. It's important to use an airlock lid for this process. Using an airtight lid isn't recommended, as it could lead to issues. If you do use an airtight lid, make sure to periodically release the pressure inside the jar to prevent unwanted bacterial and fungal growth, which could cause the sauerkraut to fail.

5. Place the jar in a room-temperature environment for about 2 weeks. During this time, the cabbage will naturally ferment and develop a tangy and crispy flavor.

 Note

I highly recommend using a **Pickle Pipe for Wide Mouth Jar** which you can find on Amazon. This accessory is designed to fit wide-mouth mason jars perfectly and helps create an optimal environment for fermenting.

Watery Cabbage Kimchi (Yangbaechu Mul Kimchi / 양배추물김치)

Total Time: 1 hour 30 minutes

Ingredients:

- ¼ cabbage (1 pound)
- ½ of medium Korean radish (½ pound)
- 3 tablespoons of coarse sea salt
- ¼ medium carrot, thinly cut into 1-inch x 1-inch pieces (1 ounce)
- ¼ stalk of scallion, sliced into 2-inch pieces (¼ cup)
- 3 cloves of garlic, sliced
- ⅛ bunch of dropwort, sliced into 2-inch pieces (2 ounces)
- ¼ of red pepper thinly sliced (1 ounce)
- 1 large pear thinly cut into 1-inch x 1-inch pieces (1 cup)
- 1 large apple half sliced, half whole (1 cup)
- 4 quarts of water (for making a 1% salt solution)

Method:

1. Clean the cabbage and radish and cut them into thin 1-inch x 1-inch pieces. Sprinkle 3 tablespoons of coarse sea salt over them and mix well. Set this aside.

2. Prepare the carrot, garlic, scallion, dropwort, red pepper, pear, and apple as directed in the

ingredient list.

3. In a large container, combine the salted cabbage and radish mixture (from step 1) with the rest of the vegetables and fruits (from step 2).

4. Pour 4 quarts of water over the ingredients in the container. Cover the container with a lid.

5. Leave the container at room temperature for one to two days to allow the fermentation process to take place.

Keep in mind that the saltiness for most watery kimchi should be about 1% of the total weight or amount of water used. Enjoy your homemade watery kimchi!

Watery Cucumber Kimchi (Oi Mul Kimchi / 오이물김치)

Total Time: 2 hours 30 minutes

Ingredients:
Brining Cucumbers:
- 5 long, thin-shaped cucumbers (2 pounds)
- 1 tablespoon of coarse sea salt
- 1 pint of water

Cucumber Stuffing:
- ¼ of radish (3 ounces)
- ¼ of purple onion (2 ounces)
- 3 cloves of garlic
- ½-inch piece of ginger
- ¼ stalk of big green onion (2 ounces)
- 2 raw shiitake mushroom
  (or ¼ oz dry shiitake) (2 ounces)
- 2 teaspoons of shrimp sauce

Watery Part:
- ⅓ of a big pear (4 ounces)
- 1 cup of cooked rice water
- 1 quart of water
- ½ teaspoon of Kosher salt
  (+ 1% saltiness shrimp sauce)

Method:

1. In a 9"x12"x1.5" cuboid container, dissolve 1 tbsp of coarse sea salt in 1 pint of water. Immerse the cucumbers in a single layer and leave them for about 2 hours. Then, wash the cucumbers in fast-running water, drain, and set aside.

2. After draining, cut off both ends of the cucumbers and slice them very thinly. Cut each cucumber piece vertically into four, leaving ¼-inch of one end.

3. Slice the radish, onion, garlic, raw shiitake, and big green onion very thinly. Place ginger slices in a cotton pouch and leave it at the bottom of the container.

4. Mix the sliced ingredients from step 3 with the cucumber end-slices and shrimp sauce. Stuff this mixture into the openings of the cucumber pieces, ensuring it's not too salty. Leave the remaining stuffing at the bottom of the container beside the stuffed cucumbers.

5. Mix water with cooked rice water and salt. Pour this mixture over the stuffed cucumber pieces, resembling rain. It's okay if the cucumber slices are floating. Add thick slices of pear on top of the water.

6. After about a day, the fermentation process will begin. Keep the container in the refrigerator and enjoy the crispy and crunchy cucumbers.

 Note

The saltiness of sea water is around 3.1% to 3.5%. In the past, Korean ancestors used this brining method to harden surfaces of vegetables like cabbages, small radishes, and cucumbers with salt in sea water. They didn't cut these vegetables but brined them, then washed away all the salty sea water, resulting in less salty kimchi. This method respects the wisdom of our ancestors.

Enjoy your delicious cucumbers! If you have any more recipes or need further assistance, feel free to ask.

Watery Radish Kimchi (Mu Mul Kimchi / 무물김치)

Total Time: 1 hour

Ingredients:

- 1 large Korean radish or 2 small Korean radishes (2 pounds)
- ½ medium pear
- ¼ of an apple
- ½ of small onion
- 2 stalks of scallion
- 3 cloves of garlic
- ⅓ inch piece of ginger
- 2 quarts of water
- 2 ounces of coarse sea salt or Kosher salt

Method:

1. Cut ¾ of the radish into ⅔″ x ⅔″ x ⅛″ cuboids.
2. Sprinkle coarse sea salt over the cuboid radish (#1), making sure they are well-covered, and let them sit for 30 minutes. Set them aside.
3. Cut the remaining ¼ of the radish into random pieces. Grind these radish pieces in a chopper and then squeeze out all the radish juice using a cotton or burlap pouch. Set this juice aside.
4. Chop the apple, pear, onion, garlic, and ginger, and place them all together into another cotton pouch. Set this pouch aside.
5. Fold the scallion and tie it with a piece of cooking thread. Set this aside.

6. You'll notice that the radish cuboids from step 2 have become semi-transparent. Use all the salty radish juice that was released during the salting process. Place the salted radish cuboids into a container, cover them with the radish juice from step 3 and the pouch of fruits and vegetables from step 4. On top of this mixture, pour 2 quarts of water. If the mixture isn't salty enough, add some more salt.
7. Leave the "watery radish kimchi" at room temperature. After one day, you should see bubbles forming on the surface, indicating that fermentation has begun. Transfer it to the refrigerator, and it will gradually develop its unique flavor.

Watery Sedum Kimchi (Dolnamul Mul Kimchi / 돌나물물김치)

Total Time: 1 hour 30 minutes

Ingredients:

- 1.6 pounds of sedum (cleaned and washed)
- ¼ of small Korean radish, cut into 1-inch x 1-inch squares (½ pound)
- 6 tablespoons of coarse sea salt
- (2 tablespoons for radish salting + 4 tablespoons for kimchi liquid)
- ½ stalk of big green onion
- 3 cloves of garlic, diced
- ⅖ -inch piece of ginger, diced (⅕ ounce)
- 1 large pear half ground, half at the bottom
- ¼ red pepper, chopped (1 ounce)
- 12 quarts of water
- ½ cup of multi-grain rice (grinded, in a pouch)

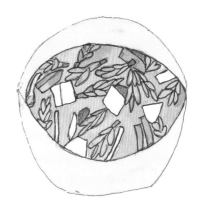

Method:

1. Clean and wash the sedum about 10 times, then drain. Set it aside.

2. Sprinkle 2 tablespoons of coarse sea salt on the cut radish and wait for 30 minutes. Do not wash the salted radish and save the juice.

3. In a large container, prepare 12 quarts of water and add 4 tablespoons of salt, including the radish juice from step 2. Mix well. Add the salted radish, big green onion, garlic, ginger, pear, red pepper, and ground rice into the water.

4. Place the cleaned sedum at the bottom of the container. Pour the liquid mixture from step 3 over the sedum.

5. Cover the container with a lid and leave it at room temperature for about 2 days. The watery sedum kimchi will be quite salty at this point (3% saltiness). When serving, dilute it with two times more water than kimchi liquid (1% saltiness) and add sugar as a taste enhancer.

White Kimchi (Baek Kimchi / 백김치)

Total Time: 4 hours

Ingredients:

- 1 Napa cabbage (2 pounds)
- ½ cup of coarse sea salt
- ½ of a large pear
- ⅛ bunch of red leaf mustard (2 ounces)
- ¼ stalk of big green onion
- 10 chestnuts, sliced (3 ounces)
- 2 sheets of rock tripe mushroom (¼ ounce)
- 2 cloves of garlic, sliced
- ⅓ inch piece of ginger, sliced
- half of red pepper, stir-fried without oil (1 ounce)
- 2 teaspoons of shrimp sauce
- 2 big shrimps (2 ounces)
- ½ pint of water

Method:

Quick Salting:

1. Choose good quality Napa cabbage.

2. Cut the Napa cabbage into quarters or halves, depending on its size.

3. In a large bowl, sprinkle coarse sea salt between the cabbage leaves, paying extra attention to the thicker stem parts. Ensure that each leaf is lightly salted.

4. Allow the cabbage to sit for about 2-3 hours. Turn the cabbage every 30 minutes so that it's evenly salted.

5. Under cold running water, wash cabbage vigorously. Drain and set it aside.

Kimchi Preparation:

6. Peel the skin of the pear and slice two-thirds of it.

7. Slice two-thirds of the white part of the big

green onion, and chop the green part.

8. Slice two-thirds of the stalk of the red leaf mustard sideways and chop the leafy part.

9. Open and peel the chestnut, then slice it. You can buy peeled chestnuts from the market to save time and effort.

10. Blanch the rock tripe mushroom, rub it with your hands to remove any dirt, wash it thoroughly, and then slice it.

11. Slice the garlic and ginger thinly, making sure they are not too thick.

12. In a big bowl, combine the sliced pear, chopped red leaf mustard, sliced chestnut, sliced rock tripe mushroom, sliced garlic, and sliced ginger. Season this mixture with shrimp sauce, but don't make it too salty as you'll season it further later.

13. Insert and pack all of the filling from step 7 between the leaves of the cabbage.

Assembly:

14. On the bottom of a container, place one-third of the red leaf mustard and one-third of the sliced pear. Add the raw shrimp on top.

15. Stack the prepared cabbage kimchi (from step 7) on top of the ingredients in the container.

16. Dissolve the shrimp sauce in water and pour it over the kimchi until it's thoroughly moistened, as if it's raining.

17. Cover the top of the kimchi with the remaining leafy cabbage leaves, press it down with a heavy object, and cover the container with a lid.

18. Leave the container at room temperature until the fermentation process starts. Once it has reached your preferred taste, store it in the refrigerator. Keep in mind that white kimchi tends to lose its taste rather quickly, so consume it promptly when it's at its best!

Whole Kimchi (Baechu Kimchi / 배추김치)

Total Time: 3 hours 30 minutes

Ingredients:

- 2 medium-sized Napa cabbages (4 pounds)
- ⅔ medium-sized radish (1 pound)
- 2 cups of coarse sea salt
- 8 quarts of water

<calculation for **quick** brine>

>1 quart = 1 liter, 3% = 30 grams, 30grams x 8 quarts = 240 grams
>240gram x **2** => 16 ounces of salt= 2 cups of salt

- 4 cloves of garlic
- ⅔ inch piece of ginger
- ¼ of big green onion
- ¼ of onion
- ⅔ cup of red pepper flake
- ⅓ cup of shrimp sauce
- 8 raw shrimps (⅔ cup)
- ⅔ large fuji apple
- ⅓ large pear

Method:

1. Choose Napa cabbage: It should be stout and short, with soft green leaves covering the outside. Cut the root part about 2 inches deep and open it with both hands, preventing the leaves from breaking. Make another 2-inch cut on the root part of each half of the cabbage. Set all four cabbage pieces aside.

2. Sprinkle more salt on the white part and less salt on the leafy parts of the cabbage. Ensure each layer of cabbage stalk and leaves is salted. Place them in a wide container. Dissolve the remaining salt in water and pour it over all four cabbage pieces.

3. After an hour, flip the cabbage pieces over and wait for 20-30 minutes. Using double, the salt than needed for brining allows for a shorter brining time while achieving a less salty, yet crispier and juicier result.

4. Hold the root part of the cabbage and shake it vigorously under fast-running water, as if you're mad at the cabbage! Do this for just one wash. Place them on a wide basket to drain.

5. Slice the radish into ⅛ -inch thickness.

6. Use a food processor to grind garlic, ginger, onion, apple, pear, and raw shrimp. Slice the white part of the big green onion and chop the green leafy part. Set aside.

7. In a large bowl, mix #5, #6, and red pepper flake. Fill the cabbage layers with the mixture

from #7 and place them inside a kimchi container. Let them ferment at room temperature for about a day until fermentation starts. Then, store the kimchi in the refrigerator. Cut into 3-inch lengths and serve.

Kimchi Science:
Coarse sea salt helps harden the cabbage's surface, retaining cabbage juice during the salting process. The main concern in kimchi-making is preserving maximum cabbage juice. Sea salt's saltiness is 3-3.5%, and Korean ancestors wisely used sea water for making kimchi. Fine salt can penetrate the outer cabbage leaves and extract juice, resulting in a salty and bitter taste.

Circumstances vary, like thin leaves in spring and thick leaves in autumn, room temperature changes, or excessively long brining times. Onion inhibits harmful bacteria growth in kimchi.

Harmful bacteria thrive on white sugar. Adding white sugar while making kimchi can cause bad bacteria to outgrow kimchi Lactobacilli. Kimchi Lactobacilli prefer fructose in fruits and thrive on fruit extracts.

Shrimp sauce is not for us, but a source of amino acids for kimchi Lactobacilli. Rice serves as a carbohydrate source for them.

Traditionally, Koreans buried kimchi in breathable earthen pots. Natural circulation due to weather changes caused kimchi juice to go down on cold days and up on warm days, enhancing the kimchi's flavor.

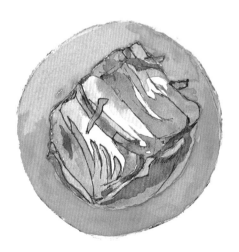

Whole Radish Kimchi (Chonggak Kimchi / 총각김치)

Brining Time: 10 hours
Seasoning Time: 2 hours

Ingredients:

- 2 bundles of whole radish (6 pounds)
- ¾ cup coarse sea salt
- 6 quarts water
- 2 tablespoons red pepper flakes
- 1 package of red pepper (5 ounces)
- ½ stalk of big green onion
- 6 cloves of garlic
- 1-inch piece of ginger
- ½ of a large pear
- ½ of a large apple
- ½ tablespoon shrimp sauce
- 1 ½ tablespoons anchovy extract
- ½ cup rice paste
  (rice powder 2 tablespoons + water ½ cup)

Method:

1. When you purchase bundles of whole radishes from the market, there's no need to clean or trim them. Simply mix the sea salt in the water and brine the radishes in the solution for 8-12 hours.

2. Rinse the radishes thoroughly under a generous flow of water. If your washing container has a large opening at the top, it's helpful for flushing out the saltwater that clings to the radish's surface. Additionally, ensure you clean the area between the stem and the radish by gently removing any traces of soil. Rinse once more and allow them to drain. For larger radishes, consider splitting them in half. Set them aside.

3. Dissolve rice powder in water to create rice paste by stirring, or alternatively, you can use cooked rice. Allow it to cool, then set it aside.

4. In a large food processor, finely chop and grind pear, apple, red pepper flakes, red pepper, garlic, ginger, anchovy extract, shrimp sauce,

and rice paste. Set this mixture aside.

5. Thinly slice the white part of the large green onion and chop the green leafy portion.

6. In a spacious bowl, combine the soaked whole radishes from step 3, the mixture from step 4, and the sliced green onions from step 5. It's worth noting that this recipe omits sugar but includes a portion of fruits. @ Chonggak kimchi is also known as ponytail radish kimchi.

Young Radish Kimchi (Yeolmu Kimchi / 열무김치)

Brining Time: 4 hours
Seasoning Time: 2 hours

Ingredients:

- 3 bundles of young radish, leaves with root (3 pounds)
- 4 quarts of water
- ½ cup of coarse sea salt
- 2 tablespoons of red pepper flakes
- ⅜ stalk of big green onion
- ½ of sweet onion, sliced
- 3 cloves of garlic, minced
- ½-inch piece of ginger, minced
- ½ of green pepper, chopped
- 1 red pepper, minced (1 ounce)
- 1 whole apple, sliced
- 1 tablespoon of shrimp sauce
- ½ cup of flour
- ¼ cup of water

Method:

1. Begin by trimming any yellowish leaves from the young radish bunches. Brine the young radish leaves with roots in 4 quarts of water and ½ cup of coarse sea salt for about 4 to 6 hours. Allow them to soak in the salty water.

2. After brining, wash the young radish leaves thoroughly under a large amount of fast-running water. Use a container with a wide opening to wash them, which helps remove the salty water from the leaves' surface. Peel off any traces of soil between the stems and roots. Wash the leaves again and drain them. Lay the drained leaves down straight and cut them into 3-inch lengths. Set them aside.

3. Dissolve the flour into the water and boil it to create a flour paste. Let the paste cool, and then set it aside.

4. Slice the white part of the big green onion and

chop the green parts.

5. In a wide bowl, combine the flour paste (#3), sliced white part of the big green onion (#4), red pepper flakes, minced garlic, minced ginger, sliced sweet onion, chopped green pepper, minced red pepper, sliced apple, and shrimp sauce. Mix all these ingredients well.

6. Add the drained young radish leaves (#2) to the mixture in the bowl. Gently fold the ingredients together; avoid rubbing the young radish leaves too hard. Transfer the mixture to a container.

7. Allow the mixture to sit at room temperature for one to two days.

Once the fermenting process is complete, you can enjoy this flavorful and spicy kimchi made from young radish leaves. It makes for a delightful addition to your meals.

Artist: Kyung Shin
Title: Dreamy Victoria Falls
Size: 12"×9"
Medium: watercolor

Chapter VI

Desserts

Baked Sweet Potatoes in Air Fryer

Total Time: 30 minutes

Ingredients (2 persons):
- 2 medium sweet potatoes (2 pounds)

Method:

1. Choose long-shaped sweet potatoes that are neither too big nor too small. Clean and wash the outside thoroughly with running cold water. Drain and set them aside.

2. Preheat your oven to 400 degrees Fahrenheit and set a timer for 20 minutes. This should be enough to bake them fully, but if you want to recreate the childhood memory of eating sweet potatoes with slightly blackened shells and easily separable insides when peeling, consider baking them for an additional 10 to 15 minutes.

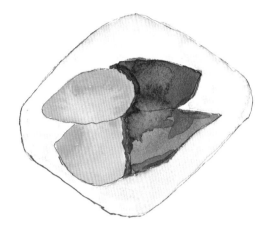

 Note

Air fryers are incredibly convenient! Once you become proficient in using one, you'll have no issues. Pair these sweet potatoes with fully fermented kimchi for a perfect snack. Keep in mind that it typically takes 45 minutes to bake sweet potatoes in a conventional oven.

Bean Juice (Kongguk / 콩국)

Soaking Time: overnight
Preparation Time: 30 minutes

Ingredients (6 persons):
- 2 cups of black beans (4.4 ounces/125 grams)
- 2 tablespoons of roasted perilla powder
- water (twice the amount of swollen beans)

Method:

1. Wash and clean the black beans, then immerse them in cold, clean water overnight.
2. The next morning, add some more water to the soaked beans and boil them for 12 minutes. The shells of the beans may float to the surface; you can discard them, but some people prefer to grind them together with the cooked beans as it enhances the flavor. When boiling, be cautious as it can potentially overflow, so keep the lid slightly open.
3. Do not discard the water used for boiling. Let it cool, and then add 2 cups of water. Blend the beans and water in a blender along with the roasted perilla powder until well mixed.

4. Your black bean perilla drink is now ready to be served. You can store it in the refrigerator.

Enjoy it as a refreshing drink or use it as a base for bean noodle soup.

Carrot Patties

Total Time: 30 minutes

Ingredients (4 children):
- 2~3 medium carrots (5 ounces)
- ¼ teaspoon of Kosher salt
- 2 tablespoons of potato starch
- 1 tablespoon of grapeseed oil

Method:

1. Slice the carrot very finely, as thinly as possible.

2. In a mixing bowl, combine the carrot slices, potato starch, and salt. Allow the mixture to sit for 5 minutes. The moisture from the carrot slices will be enough to create a batter for the carrot patties.

3. Heat a pan and add grapeseed oil, swirling it around to coat the pan. Stir-fry the carrot mixture until the edges turn brown and it begins to emit the aroma of roasted sweet potatoes. Flip it over just once.

4. Serve the carrot patties.

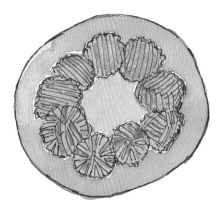

 Note

These patties have a crispy and sweet taste that may surprise you, making them a delightful and healthy snack. Children, even those who don't typically enjoy carrots, tend to love them. They're highly recommended as a nutritious snack for kids.

Cinnamon Ginger Punch (Sujeongwa / 수정과)

This recipe is for making a traditional Korean ginger and cinnamon punch. It's a sweet and spicy beverage with the flavors of ginger and cinnamon. Here's how to make it.

Total Time: 2 hours

Ingredients (6-8 persons):
- 30 cinnamon sticks (4 ounces)
- ¾ cup of ginger, sliced (3 ounces)
- 5 ounces of organic sugar (or adjust to taste)
- 1 tablespoon of pine nuts
- 4 quarts of water
- 6 ounces of dried persimmon (3 persimmons) (optional)

Method:

1. Clean the ginger, peel it, and slice it. Soak for about 15 minutes to remove excess starch. Then, drain and set them aside.

2. Brush the cinnamon sticks thoroughly, inside and outside under running water. Set them aside.

3. In a medium pot, start to boil 2 quarts of water with the # 1 sliced ginger. Once it reaches a boil, reduce the heat to medium and let it simmer for I hour.

4. In another medium pot, start to boil 2 quarts of water with the # 3 cleaned cinnamon sticks. Once it boils, remove any bubbles that form, and let it simmer for 1 hour.

5. From each pot, take out the ginger slices from

one and the cinnamon sticks from the other.

6. In a larger pot, combine the # 3 ginger-infused liquid and the # 4 cinnamon-infused liquid. Add sugar to taste and bring the mixture to a boil. Remove any extra bubbles that form during boiling.

7. Allow the ginger and cinnamon punch to cool, and then refrigerate it. Optionally, you can cut the dried persimmon into pieces and immerse it into the punch in a separate container. Serve the ginger and cinnamon punch chilled with pine nuts as a garnish.

 Notes

• It is important to boil the ginger and cinnamon separately to prevent them from affecting each other's flavors. Combining them later creates the perfect balance of spiciness from ginger and warmth from cinnamon.

• Dried persimmon can be added as a sweet addition if desired.

Cinnamon Spiced Sweet Potatoes

Total Time: 40 minutes

Ingredients (4~6 persons):
- 3 medium sweet potatoes, peeled and cut in half lengthwise (3 pounds)
- melted salted butter
- 2 tablespoons organic sugar
- ½ tablespoon soy sauce
- ½ teaspoon ground cinnamon

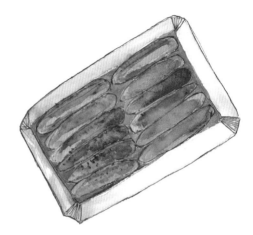

Method:

1. In a large pot, parboil the sweet potatoes until they are about halfway cooked. This will ensure they have a soft texture after baking. Once done, peel off the skin and cut them in half lengthwise.

2. Preheat your oven to 350-degree Fahrenheit. Place a baking sheet on a baking tray and evenly arrange the sweet potato halves on it. Brush the tops of the sweet potatoes with half of the melted butter.

3. In a small bowl, mix the organic sugar, soy sauce, and ground cinnamon. Sprinkle half of this mixture over the sweet potatoes.

4. Bake the sweet potatoes in the preheated oven for 15 minutes. After that, flip them over, brush the other side with the remaining melted butter, and sprinkle the rest of the cinnamon mixture over them. Continue baking for an additional 15 minutes or until the sweet potatoes are tender.

Cranberry Sauce

Total Time: 1 hour

Ingredients (4 persons):
- 1 bag of cranberries (¾ cup)
- ¾ cup of orange juice
- 1 tablespoon of lemon juice
- 1 cup of organic sugar
- 1 teaspoon of orange zest
- ⅛ teaspoon of cayenne pepper

Method:

1. First, remove any spoiled cranberries, then wash the remaining cranberries in cold running water. Drain them and set them aside.

2. In a medium-sized saucepan, combine the washed cranberries, orange juice, lemon juice, organic sugar, orange zest, and cayenne pepper. Initially, heat the mixture over high heat for 2 minutes, then reduce the heat to medium. Continue stirring for 10 to 15 minutes. Let it cool before serving.

Cream Puffs

Total Time: 1 hour

Ingredients (Makes 20 pieces):
- 3 tablespoons of unsalted butter
- ¼ teaspoon of Kosher salt
- ¼ cup of water
- 3 tablespoons of milk
- ¼ cup of cake flour
- 4 ounces of egg (equivalent to 2 eggs)
- 3 tablespoons of powder sugar (for later)

Method:

1. Place the unsalted butter, salt, milk, and water in a pot and bring them to a gentle boil over low heat. Set it aside.

2. Sift the cake flour twice and combine it with the mixture from step 1. Simmer and stir the mixture over low heat until it turns transparent.

3. Transfer the mixture from step 2 into a bowl and allow it to cool.

4. Whisk the eggs and gradually pour them into the mixture from step 3, stirring until you have a somewhat thick batter.

5. Line a rectangular pan with wax paper and drop ½ tablespoon of the batter from step 4 to form 20 individual pieces.

6. Bake for the first 15 minutes at 370 degrees Fahrenheit, then reduce the temperature to 340 degrees Fahrenheit and bake for an additional 10 minutes.

7. Be sure not to open the oven door during the baking process, as opening it prematurely can cause the puffs to deflate.

Deep-fried & Sugar Glazed Banana

Total Time: 1 hour

Ingredients (4~6 persons):
- 3 bananas (approximately 1 pound)
- 2 tablespoons of wheat flour
  (for coating sliced banana)
- 2 cups of grapeseed oil (for frying)

Batter/Coating:
- 2 eggs (4 ounces of egg)
- 3 tablespoons of wheat flour
- 3 tablespoons of cornstarch
- 2 tablespoons of water

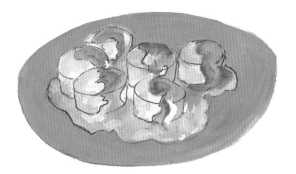

Syrup Making:
- 1 tablespoon of grapeseed oil
- ¾ cup of organic sugar

Method:

1. Slice the bananas into 1-inch-thick rounds and coat them thoroughly with wheat flour.

2. Prepare the batter/coating by mixing the eggs, wheat flour, corn starch, and water. Whisk them together until well combined.

3. Coat the banana slices (prepared in step 1) with the batter from step 2, ensuring they are well coated.

4. Heat the grapeseed oil in a deep-frying pan or pot. Once the oil is hot, carefully add the coated banana slices and fry them until they are golden brown and crispy. Remove and drain on paper towels.

5. To make the syrup, heat 1 tablespoon of grapeseed oil and ¾ cup of organic sugar in a separate pan. Watch it closely without stirring until the sugar changes color to a transparent golden yellow. To test if it's done, take a small amount on a ladle and see if it produces soft and fine threads when dripped.

6. Mix the syrup from step 5 with the fried banana slices from step 4. Roll them in a cold-water puddle to help separate them and prevent them from sticking together.
7. Serve the deep-fried and sugar-glazed banana slices on a pretty plate.

This recipe aims to create a sweet and crispy treat.

Deep-fried & Sugar Glazed Sweet Corn Ball

Total Time: 1 hour

Ingredients (4~6 persons):
- 2 cups of sweet corn kernels
- 3 tablespoons of wheat flour
- 3 tablespoons of cornstarch
- 1 egg
- ⅛ teaspoon of Kosher salt
- 2 cups of grapeseed oil (for frying)
- ¼ cup of cold water (for later)

For the syrup:
- 1 tablespoon of grapeseed oil
- ¾ cup of organic sugar

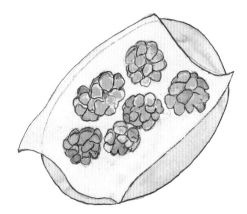

Method:

1. Mash half of the sweet corn kernels and add the egg, wheat flour, and corn starch to it. Mix everything well. Don't add more flour even if the batter seems thin. You can add a bit more egg if needed.

2. In a deep-frying pan, heat 2 cups of grapeseed oil. Once it's hot, spoon out one tablespoon of the corn mixture (#1) at a time and fry them one by one until they turn a golden yellow color. Drain the excess oil and set the fried corn aside.

3. To make the syrup, heat 1 tablespoon of grapeseed oil in a separate pan, then add the organic sugar. Do not mix them at the beginning. Watch the mixture closely, and once the sugar has melted completely and starts changing color to golden brown without bubbling, take some of it out with a ladle. If you see hardened threads forming in the sugar, it's time to stop boiling.

4. Mix the fried sweet corn (#2) with the syrup you've prepared in step 3. Roll each piece in a shallow bowl with cold water or oil to prevent them from sticking together. Then, serve them on a plate.

Notes

• Be cautious when making the syrup, as it can quickly turn into caramel if you overcook it.

• This dish can be served as a dessert and is likely to be enjoyed by children.

It's important to follow the steps carefully, especially when making the syrup, to achieve the desired results. Enjoy deep-fried & sugar-glazed sweet corn balls!

Fruit Compote / Fruit Nut & Rice Salad

Total Time: 1 hour

Ingredients (4~6 persons):
- 1 bag of Uncle Ben wild long rice (2 cups)
- 4 cups of boiling water
- 2 tablespoons of Dijon Mustard
- ⅓ cup of raspberry vinegar
- ⅔ cup of extra virgin olive oil
- ¼ large yellow bell pepper, chopped (¼ cup)
- ¼ large red bell pepper, chopped (¼ cup)
- 1 cup of sliced scallions
- 1 cup of dry cranberries, cherries, and raisins (total)
- 1 cup of roasted pecans
- 8 long sprigs of parsley, chopped (⅓ cup)

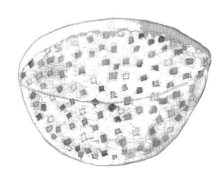

Method:

1. In a pot, add rice, 1 tablespoon of oil, and boiling water. Cover with a lid and bring to a boil. Once it boils, reduce heat to low, and simmer for about 30 minutes. Remove the pot from the heat, and let it sit for another 30 minutes. Cool and set aside.

2. In a large bowl, whisk together Dijon Mustard, raspberry vinegar, and the remaining olive oil until well combined.

3. To the marinade in step 2, add the yellow and red bell peppers, scallions, dried cranberries, cherries, raisins, chopped pecans, and the cooled rice. Mix everything together to coat evenly. Sprinkle chopped parsley on top.

4. Allow the salad to sit for at least one hour before serving. It can be kept in the refrigerator for 2-3 days.

Enjoy your Wild Rice Salad with Dijon Mustard and Raspberry Vinaigrette!

Glutinous Rice Cake (Injeolmi / 인절미)

Total Time: 2 hours

Ingredients (4~6 persons):

- 2 ½ cups of sweet rice
- ½ tablespoon of Kosher salt
- 2 tablespoons of organic sugar
- ½ cup of roasted soybean powder
- ¼ cup of roasted and crushed black sesame seed powder
- 5 quarts of water (for spraying + steaming)

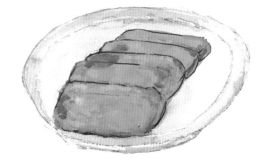

Method:

Prepare the Sweet Rice:

1. Clean and wash the sweet rice. Immerse it in cold water overnight to allow it to swell. Drain the rice using a basket.

Steam the Sweet Rice:

2. In a 6-quart pot, pour water to about ⅔ full and bring it to a boil.

3. In a 2-layered bamboo steamer, spread out a damp burlap cloth and place the swollen sweet rice on each layer.

4. Place the bamboo steamer on top of the boiling water pot. Steam the sweet rice for 30 minutes.

Pound and Mix the Sweet Rice:

5. Meanwhile, melt ½ tablespoon of salt into ½ cup of water.

6. Remove the bamboo steamer from the pot and place it on the countertop. Open the cover and spray the saltwater mixture (#3) over the rice. Mix the rice by flipping it over with a rice spatula.

7. Add some more water into the pot, bring it to a boil again, then place the bamboo steamer on top, and steam for an additional 30 minutes.

Prepare the Sweet Rice Cakes:

8. Pour out the fully steamed sweet rice onto a mortar and pound it by striking it with a pestle that is coated with saltwater. Continue pounding until the rice is fully crushed and a stickiness develops.

9. Add sugar to the pounded rice and pound it by striking again. This step is crucial for achieving a good result.

Shape and Coat:

10. Wet your hands with cold water, place the rice mixture onto a cutting board, cut it into 1.5-inch lengths, and roll each piece in a mixture of soybean powder and black sesame seed powder.

11. Cover the cutting board and knife with salt water to make it easier to cut and shape the rice cakes.

Store:

12. You can pack an adequate amount of the rice cakes in plastic wrap and store them in the freezer. To enjoy them fresh and chewy, simply thaw them at room temperature whenever you like.

Enjoy your injeolmi! These sweet rice cakes have a delightful chewy texture and a nutty flavor from the soybean and sesame powders.

Korean Sweet Dessert Pancakes (Hotteok / 호떡)

Total Time: 1 hour

Ingredients (for 8 Hotteoks):
- 1 cup of all-purpose flour
- 1 cup of sweet rice powder (mochiko)
- 1 ½ cups of 2% milk (lukewarm)
- 2 teaspoons of dry yeast
- 2 tablespoons of organic sugar
- 1 teaspoon of Kosher salt
- 1 tablespoon of grapeseed oil

For the Hotteok filling:
- ½ cup of brown sugar
- 1 teaspoon of cinnamon powder
- 2 tablespoons of chopped walnuts

Method:

1. In a mixing bowl, pour 1.5 cups of lukewarm milk. Add sugar, yeast, salt, and oil, and stir them well.

2. Add all-purpose flour and sweet rice powder into the mixture from step 1 and mix them by hand to make a dough. Leave the dough covered with a lid at room temperature to rise due to the yeast inside.

3. After one hour, the dough should double in size. Then, knead it to remove any gas bubbles inside. Let it rise again for about 10-20 minutes.

4. To make the filling, mix brown sugar, cinnamon, and chopped walnuts in a small bowl. Set it aside.

5. Knead the dough again to remove any remaining gas bubbles.

6. Sprinkle some flour on your cutting board. Place the dough on the cutting board, shape it into a lump, and cut it into 8 equal pieces to form balls. Insert the filling from step 4 into the center of each ball and seal it. Make 8 balls with the filling inside and cover your palms and fingers with flour to prevent sticking.

7. Heat up your frying pan over medium heat, add some oil, and place one ball in the pan. Cook it for about 30 seconds. When the bottom turns light brown, flip it over and press the ball with a spatula to shape it into a thin, wide Hotteok. Cook for 1 minute until the bottom becomes golden brown.

8. Flip it over again and cook for an additional 1 minute, then cover the pan with a lid. The brown sugar granules will melt and turn into syrup at this point. Serve the Hotteok while they are hot.

Korean Sweet Potato Snack (Matang / 마탕)

Total Time: 1 hour

Ingredients (4 persons):
- 1 medium sweet potato (approximately 1 pound)
- 2 ½ tablespoons of organic sugar
- 2 ½ tablespoons of water
- 2 ½ tablespoons of honey
- 2 cups of frying oil

Method:

1. Begin by cutting the sweet potato into random-shaped pieces. Immerse these pieces in water for about 10 minutes. This step helps extract the starch from the sweet potato. Afterward, pat them dry with a paper towel.

2. Place the sweet potato pieces into a gallon-sized Ziploc bag, seal the bag, and microwave the pieces for 3 minutes.

3. Once microwaved, pat them dry again with a paper towel.

4. In a large pan, combine the organic sugar, honey, and water to create a glaze. Start heating the mixture, and let it come to a boil.

5. As the sugar melts and the mixture boils, carefully add the sweet potato pieces to the glaze. Sprinkle black sesame seeds on top of the glazed sweet potato pieces.

This delightful Korean-style snack requires only a few simple ingredients. If you have kids, why not surprise them with this homemade treat?

Korean Sweet Rice Cake with Dried Fruit and Nuts (Yak Sik / 약식)

Total Time: 2 hours

Ingredients (8 persons):
- 4 cups of glutinous rice
- ¼ cup of jujube or raisins
- 2 cups of peeled chestnuts (cut in half)
- ¼ cup of pine nuts
- 2 teaspoons of cinnamon powder
- ¼ cup of soy sauce
- ¾ cup of organic sugar
- ¼ cup of honey
- 3 tablespoons of roasted sesame oil
- 1 ½ cups of water

Method:

1. Wash and clean the glutinous rice, then soak it in clean water for about 2-3 hours. Drain and set aside.

2. In a mixing bowl, combine the glutinous rice, soy sauce, sugar, honey, and cinnamon powder. Mix well. Place this mixture on the bottom of the pressure cooker pot.

3. Add all the chestnuts and either jujube or raisins on top of the rice mixture. Add half of the pine nuts and ½ of the sesame oil.

4. Cook in the pressure cooker on a multi-grain setting for 40 minutes.

5. After cooking, transfer the mixture to a large mixing bowl. Add the remaining sesame oil and mix well.

6. Decorate the surface with the remaining half of the pine nuts.

This dish sounds delicious and is perfect for enjoying the unique flavor combination of chestnuts, jujube (or raisins), and glutinous rice.

Mochi

Total Time: 2 hours

Ingredients (6~8 persons):
- 1 box of Mochiko (2 cups of sweet rice flour, 16 oz)
- ½ cup of white sugar
- 2 teaspoons of Kosher salt
- 1 ½ cups of water
- 1 cup or more corn starch
- 12 ounces of sweetened red bean (⅘ of a can)

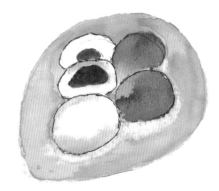

Method:

1. In a large bowl, mix the sweet rice flour, sugar, salt, and water well until you have a sticky and soggy dough. Divide the dough into two equal portions and set them aside.

2. Spread damp burlap cloth on each shelf of a bamboo steamer and place one portion of the dough on each shelf. Set it aside.

3. Boil ⅔ of water in a large pot. When the water starts boiling, stack the two shelves of the bamboo steamer on top of the pot, cover the bamboo steamer with its lid, and steam for 30 minutes without opening the lid.

4. Spread a generous amount of corn starch on a large and wide cutting board. Prepare a knife and a plate with a dusting of corn starch.

5. It's time to make the mochi! Flip the top shelf of the steamer over onto the cutting board. In other words, flip the steamed dough to cover all sides with corn starch, make it wider, and cut the dough into eight pieces. Repeat the same process for the dough on the bottom shelf. You should now have 16 pieces of mochi ready on your cutting board.

6. Shape each mochi piece into a square, place 1.5 tablespoons of sweetened red bean as a filling in the center, fold two opposite corners to meet diagonally, and then fold the other two opposite corners in the same way. Pinch the edges to seal the mochi and let them rest with the sealed side down on a plate. They will naturally take on a pleasing shape.

7. Repeat this process for the remaining 15 mochi pieces, making sure they don't touch each other on the plate. Arrange them neatly. Serve.

Namagashi (Saenggwaja / 생과자)

Total Time: 2 hours

Ingredients (6~8 persons):

- 1 box of Mochiko (16 ounce) or sweet rice flour (2 cups)
- ½ cup of white sugar
- 2 teaspoons of Kosher salt
- 1 ½ cups of water
- 1 cup or more corn starch
- 12 ounces of sweetened red bean (⅘ of a can)

Method:

1. In a large bowl, mix the sweet rice flour, sugar, salt, and water well until you have a sticky and soggy dough. Divide the dough into two equal portions and set them aside.

2. Spread damp burlap cloth on each shelf of a bamboo steamer and place one portion of the dough on each shelf. Set it aside.

3. Boil ⅔ of water in a large pot. When the water starts boiling, stack the two shelves of the bamboo steamer on top of the pot, cover the bamboo steamer with its lid, and steam for 30 minutes without opening the lid.

4. Spread a generous amount of corn starch on a large and wide cutting board. Prepare a knife and a plate with a dusting of corn starch.

5. It's time to make the mochi! Flip the top shelf of the steamer over onto the cutting board. In other words, flip the steamed dough to cover all sides with corn starch, make it wider, and cut the dough into eight pieces. Repeat the same process for the dough on the bottom shelf. You should now have 16 pieces of mochi ready on your cutting board.

6. Shape each mochi piece into a square, place 1.5 tablespoons of sweetened red bean as a filling in the center, fold two opposite corners to meet diagonally, and then fold the other two opposite corners in the same way. Pinch

the edges to seal the mochi and let them rest with the sealed side down on a plate. They will naturally take on a pleasing shape.

7. Repeat this process for the remaining 15 mochi pieces, making sure they don't touch each other on the plate. Arrange them neatly. Serve.

Organic Green Smoothie (Yachae Smoothie / 야채스무디)

Total Time: 20 minutes

Ingredients (for 4 persons):
- ⅛ of a bundle of kale (1 ounce)
- one leaf of collard greens (⅗ ounce)
- ½ of a medium carrot (2 ½ ounces)
- 1 banana (approximately 6 ounces)
- ½ of a whole avocado (5 ounces)
- 3 cups of whole milk
- 1 ½ cups of orange juice

Method:

Ensure that all your ingredients are organic.

1. You'll need a high-quality blender like the Vita-Mix brand for this recipe.

2. Begin by adding the kale and collard greens to the blender.

3. Follow with the carrots, banana, and avocado.

4. Pour in the whole milk.

5. Add the orange juice to the mix.

6. Blend all the ingredients together until you achieve a smooth and creamy consistency.

7. Pour the green smoothie into glasses and serve. This nutritious and delicious smoothie is perfect for a family breakfast and can be enjoyed by both kids and adults.

Feel free to adjust the ingredients to suit your taste and dietary preferences. Enjoy your organic green smoothie!

Pine-needle Rice Cake (Songpyeon / 송편)

Total Time: 3 hours

Ingredients (10 persons):

- 3 cups of frozen rice powder
  (from the frozen section of a Korean market)
- 1 tablespoon of Kosher salt
- ½ cup of roasted sesame seeds, crushed
- ⅓ cup of organic sugar
- ½ cup of chestnuts
  (peeled and steamed, then crushed)
- ½ tablespoon of cinnamon powder
- ½ cup of sliced jujubes
- ½ tablespoon of roasted sesame oil
- pine needles
  (from your garden or neighborhood)

Method:

1. Sprinkle salt over the rice powder and rub it with your hands. Pass it through a sieve three times.

2. Mix the salted rice powder with boiling water. Save ¼ of the rice powder mix nearby in case the rice dough is too watery, and use it as needed at the last moment. Knead the mixture for a long time until it forms a firm dough. Let it stand wrapped in a wet cloth.

3. Prepare the fillings: Mix crushed sesame seeds with sugar, steamed and crushed chestnut with cinnamon, and sliced jujubes.

4. Take out approximately 8 ounces of dough and roll it on a cutting board. Make a small ball out of a piece of the dough, shape it into a circle in the palm of your hand, fill it with one of the fillings from step 3, and close it, shaping it nicely. Repeat until all the dough and three fillings are used.

5. Rinse the pine needles well and drain them. Place them on the bottom of a two-layer bamboo steamer. Place the filled dough (songpyeon) on top of the pine needles. If pine needles are not

available, you can use pieces of burlap cloth.

6. Boil water in a pot, filling it about ¾ full. Place two stacks of bamboo steamers over the pot and steam for 40 minutes. Do not open the bamboo lid while steaming.

7. Once the songpyeon is cooked, rinse them with cold water, and then rinse again in a mixture of 2 cups of water and ½ tablespoon of sesame oil. Spread them out to cool.

Pizzelle Cookies

This recipe is for making Italian pizzelle cookies. Pizzelles are delicate and thin waffle-like cookies with a lovely anise flavor. Here's how you can make them:

Total Time: 30-60 minutes

Ingredients (4 persons):
- 3 eggs, beaten
- ⅔ cup of organic sugar
- ¾ cup of unsalted butter
- 1 cup of all-purpose flour
- 1 teaspoon of baking powder
- 2 teaspoon of anise extract
- 1 teaspoon of vanilla extract
- 1 tablespoon of powder sugar (optional)

Method:

1. In a mixing bowl, combine the beaten eggs, organic sugar, all-purpose flour, and baking powder.
2. In a medium saucepan, melt the unsalted butter.
3. Pour the melted butter into the mixture and blend everything together until you have a smooth pizzelle batter.
4. Preheat your pizzelle press according to the manufacturer's instructions.
5. Once the press is hot, spoon about 1~2 tablespoons of the pizzelle batter onto the center of the press.
6. Close the lid and cook until the pizzelles are golden brown, Cooking time may vary depending on your pizzelle press, but it typically takes about 30-60 seconds.
7. Carefully remove the pizzelles from the press and let them cool on a wire rack.
8. If desired, you can sprinkle powder sugar on top of the pizzelles for extra sweetness.

9. Store the pizzelles in an airtight container once they have cooled completely.

 Notes

• Pizzelle presses can vary, so be sure to follow the manufacturer's instructions for preheating and cooking times.
• Anise extract gives these cookies their traditional flavor, but you can adjust the amount to your taste.
• You can also experiment with other extracts, such as almond or lemon, for different flavors.
• Pizzelles can be served flat or rolled into cannoli-like shapes while they are still warm and pliable, then filled with sweet fillings if desired.

Enjoy your Italian pizzelle cookies! They make a delightful treat for any occasion.

Potato Pancake (Gamjajeon / 감자전)

Total Time: 30 minutes

Ingredients (4 persons):
- 2 large potatoes
- 1 teaspoon of Kosher salt
- 2 tablespoons of grapeseed oil

Method:

1. Start by peeling the potatoes. Then, grate them and add 2 teaspoons of Kosher salt. Discard the potato juice and save only the solid grated part.

2. Place a frying pan over high heat. Once the pan is heated, reduce the heat to medium-low and swirl 2 tablespoons of grapeseed oil.

3. Pour the grated potatoes onto the pan to make pancakes with a 2-inch diameter. Maintain medium-low heat until the shape of the pancake is hardened. Then, reduce the heat to low and wait for 30 seconds before flipping it over. Be patient when flipping to avoid the pancake from sticking to the pan. Continue frying.

Enjoy your delicious Korean potato pancakes! These simple yet tasty pancakes are great for snacks, especially loved by children for their soft texture and straightforward flavor.

Red Bean Sediment (Pat Anggeum / 팥앙금)

Total Time: 2 hours

Ingredients (4 persons):
- 2 cups of red beans
- 1 teaspoon of Kosher salt
- 1 cup of organic sugar (adjust to taste)
- 5 quarts of water

Method:

1. Start by boiling the clean red beans in an ample amount of water over high heat. Once they come to a boil, turn off the heat and discard the first batch of boiled water.

2. Add the 5 quarts of fresh water and continue boiling until the red beans become soft enough to easily mash with a spatula.

3. Mash the beans and strain them through a sieve to remove the shells. Discard the residue and rinse the sieve.

4. Strain the sieved liquid once more. Pour this liquid into a cloth pouch and squeeze out the water.

5. Retrieve the sediment from the pouch and cook it with sugar and salt. Although the sediment may seem dry initially, it will become watery once you apply heat. Therefore, keep it on the heat and stir continuously. You'll notice bubbles and splashes covering the mixture.

6. When the splashing subsides and big holes appear on the surface with steam rising, this is a sign that the paste is almost done. As the steam begins to disappear, quickly turn off the heat to prevent it from hardening upon cooling.

7. Place the paste into freezer zip-lock bags and store them in the freezer. It will maintain its freshness for several months.

 Note

This red bean sediment is perfect for use as a filling when making mochi. You have the flexibility to control the sweetness to your liking.

Rice Cake Ball (Gyeongdan / 경단)

Total Time: 2 hours

Ingredients (4 persons):
Rice Balls

- 2 cups of glutinous rice powder
- 1 cup of water
- 2 tablespoons of organic sugar
- 1 teaspoon of Kosher salt

Coating Powder

- 1 cup of roasted yellow bean powder (available from the market)
- ½ cup of crushed, roasted black sesame seeds (available from the market)
- ½ cup of pitted and minced jujube

Method:

1. Sift the glutinous rice powder, sugar, and salt together three times.

2. Boil water and add it to the sifted mixture of glutinous rice powder, sugar, and salt. It's important to have plenty of water ready because the glutinous rice powder from the freezer section might not have enough moisture. Add the boiling water gradually while kneading the mixture into dough. Aim for a consistency that is neither too soggy nor too dry. Roll the dough into balls, each about 1-inch in diameter.

3. Cook the balls in a large pot of boiling water. Initially, all the balls will sink to the bottom of the pot, but as they are cooked, they will rapidly float to the surface together. Use a soup net-ladle to retrieve them, then rinse them in cold water. Finally, roll each ball in the three-coating powder.

Rice Punch (Sikhye / 식혜)

This recipe outlines how to make a traditional Korean malt drink called "Sikhea." Sikhea is a sweet and refreshing rice beverage. Here's the method:

Total Time: 6 hours

Ingredients (4-6 persons):
- ½ cup of malt powder
- 1 ¼ cups of rice
- 5 ounces of organic sugar (or adjust to taste)
- ½ tablespoon of ginger
- 6-8 quarts of water (you'll use only the top clear water without malt sediment)
- 1 tablespoon of pine nuts

Method:

Prepare the Malt Extract:

1. Put the malt powder inside a large cotton pouch and place it in a deep bowl.

2. Pour 6-8 quarts of water over the malt pouch and knead it gently, as if you are massaging it in the water.

3. After 2 hours, slowly pour out only the top clear water, being careful not to disturb the sediment. Set this clear liquid aside.

Cook the Rice:

4. Cook the rice in an electric rice cooker using slightly less water than usual.

Combine Malt Extract and Rice:

5. Mix the cooked rice with the malt extract (step 1) in a large pot.

6. Warm this mixture up to 140 degrees Fahrenheit. It's crucial not to exceed this

temperature.

7. You can check the temperature by feeling the liquid with your hand. If it feels too warm, turn off the heat. Alternatively, use a kitchen thermometer.

Ferment:

8. Transfer the mixture to the rice cooker pot and set it to a warm setting. Allow it to ferment for about 4-6 hours.

Add Sugar and Ginger:

9. Once some rice granules start floating in the liquid, pour the mixture into a large pot.
10. Add sugar to taste and a piece of ginger.
11. Boil it briefly to dissolve the sugar and remove any bubbles that form on the surface.

Cool and Serve:

12. Allow the mixture to cool and then store it in the refrigerator.
13. Serve the Sikhea chilled with pine nuts as a garnish.

 Notes

• Maintaining the correct temperature is crucial in this process. Too high a temperature can destroy the amylase enzyme responsible for converting rice starch to sugar, and too low a temperature may lead to spoilage.
• Sikhea is a traditional Korean rice beverage that's sweet, aromatic, and slightly fizzy. It's a delightful treat, especially during hot weather.

Scones

Total Time: 45 minutes

Ingredients (4 persons):
- 2 cups of soft wheat flour (cake flour)
- ¼ cup of organic sugar
- ¼ teaspoon of Kosher salt
- 1 teaspoon of baking powder
- ¼ teaspoon of baking soda
- ½ cup of unsalted butter (1 stick), cold
- ½ cup of cranberries

Method:

Preparation:

1. Sift together the soft flour, organic sugar, Kosher salt, baking powder, and baking soda. This sifting process will help ensure even distribution of these dry ingredients.

2. Dice the cold unsalted butter into small pieces. You can take it out of the refrigerator just before you start mixing the dough.

3. Toss the cranberries into the dry mixture and mix them roughly.

4. Using a rubber spatula, mix the diced cold butter with the dry mixture. Work the butter into the flour mixture until it resembles coarse crumbs. It's okay if there are still small chunks of butter visible.

5. Make a well in the center of the mixture and pour in the milk. Stir gently with a spatula until the dough comes together. Be careful not to overmix; you want the dough to be just combined.

6. Turn the dough out onto a floured surface and gently knead it a few times, just until it holds together. Pat the dough into a circle about 1-inch thick.

7. Use a sharp knife or a round cutter to cut the dough into scone shapes. You can make them as large or small as you like.

8. Place the scones on a baking sheet lined with

parchment paper, leaving some space between each one.

9. Brush the tops of the scones with a little extra milk or cream to help them brown.

Baking:

10. Preheat your oven to 375°F (190°C).

11. Bake the scones in the preheated oven for about 15-20 minutes, or until they are golden brown on top and cooked through. The exact time may vary depending on the size of your scones, so keep an eye on them.

12. Once baked, remove the scones from the oven and let them cool on a wire rack.

Enjoy these delicious cranberry scones! They're perfect for breakfast or as a snack with a cup of tea or coffee.

Sorghum Balls
(Susu Pat Danji or Susu Gyeongdan / 수수팥단지, 수수경단)

Total Time: 2 hours

Ingredients (4 ~6 persons):
- 2 ½ cups of glutinous sorghum powder (*)
- ½ cup of glutinous rice powder (*)
- ¾ cup of boiling water
- ½ tablespoon of Kosher salt
- 1 cup of cooked and crushed red beans
- 2 tablespoons of organic sugar

(*) Available in the freezer section of a Korean market

Method:

1. In a mixing bowl, combine the glutinous sorghum powder, glutinous rice powder, and salt. Sift the mixture three times to ensure a smooth texture. Add the boiling water and knead the mixture until it reaches a dough-like consistency, similar to dumpling dough. Shape the mixture into balls and flatten the center of each one. Continue making balls in this shape until all the dough is used.

2. Boil the red beans, discard the first batch of boiled water to remove any saponin, and then add fresh water. Boil the red beans again until they become soft to the touch. Drain the beans and crush them on a cutting board. While crushing, sprinkle a little sugar and salt on the pushing rod.

3. In a pot, bring water to a boil and add the prepared balls. Initially, they will sink to the bottom of the pot, but as they cook, they will float to the surface. Use a net-ladle to gather the floated balls and place them in a wide bowl with the crushed red beans. Allow them to cool slightly. It's important not to rush the rolling process while the balls are still hot, as they may lose their shape and harden quickly. This knowledge comes from experience, though the

exact reason isn't clear.

4. After a couple of minutes, wet your hands with water and roll the balls in the crushed red beans.

 Note

On my grandson's first birthday, I had the joy of being a grandmother and making sorghum balls.

Sticky Rice Cake, Covered by Red Beans (Siru Tteok / 시루떡)

Total Time: 3 hours

Ingredients (10-12 persons):
Part I - Boiled & Crushed Red Beans:
- 1 pound of red beans
- 1 tablespoon of coarse sea salt
- 2 quarts of water

Part II - Sticky Rice Cake:
- 2 pounds of rice powder
- 1 tablespoon of fine salt
- 2 tablespoons of organic sugar
- ½ cup of water (for spraying)

Method (Part I - Boiled & Crushed Red Beans):

1. Wash and clean the red beans, removing any tiny stones. Drain and set aside.

2. In a large pot, fill it ⅔ full of water and bring it to a boil. Once boiling, add the salt and red beans, and continue boiling until the red beans are completely soft. Drain and discard most of the boiling water, leaving only a small amount. Simmer over low heat for about 15-20 minutes. Open the lid of the pot and stir well with a large wooden spatula, reaching the bottom of the pot until there is no more moisture on the surface of the red beans.

3. Once the red beans become fluffy, mash them halfway. Set them aside.

Method (Part II - Sticky Rice Cake):

4. Mix the rice powder, salt, and sugar well. Sift this mixture through a fine sieve three times.

5. Prepare two sheets of damp burlap cloth and place them on each shelf of a bamboo steamer.

On top of the burlap cloth, spread a ⅓ inch layer of red beans, followed by a 1.5-inch layer of rice powder mixture, and another ⅓ inch layer of red beans. Spray water evenly over the rice mixture and red beans. Repeat the same process on the second shelf of the bamboo steamer, creating two stacks of sticky rice cakes with layers of red beans.

6. Boil ⅔ of the water in a large pot. Once boiling, place the two shelves of the bamboo steamer stacked on top of each other on the pot and cover the bamboo steamer with its lid. Steam for 30-40 minutes without opening the lid.

7. Have two large plates ready. Using mittens on both hands, carefully place the second shelf of the bamboo steamer upside down on one of the plates. Check that the rice cake has come out of the shelf, remove the burlap cloth, and flip the cake onto the other plate. This will result in fluffy red beans showing on top, with the flat red beans at the bottom of the plate. Repeat the same process with the first shelf of the bamboo steamer.

You can make variations by using sticky glutinous rice or mixing two types of grain powder. You can also insert lettuce leaves, zucchini slices, or radish cuts between the layers of rice and red beans as desired. Enjoy your Red Bean Sticky Rice Cake!

Tiramisu / Ladyfinger Coffee Cake

Total Time: 1 hour 30 minutes

Ingredients (8 persons):
- 4 packages of ladyfingers
- 2 mugs of coffee
- small instant vanilla pudding mix
- 2 cups of 1% milk
- 2 tablespoons of unsweetened cocoa
- ¼ cup of white sugar
- ¼ cup of water
- ¼ cup of unsalted butter
- 7-8 big semi-sweet chocolate morsels
- 1 pint of heavy cream

Method:

1. Spread coffee-dipped ladyfingers in a single layer in a rectangular-shaped cake pan. Freshly brewed coffee is preferred over instant coffee. Regardless, make sure the coffee has cooled a bit before dipping the ladyfingers. If the coffee is too hot, the dipping process will be challenging.

2. Prepare the vanilla pudding. Pour the 2% milk over the instant vanilla pudding mix and whisk for 2 minutes. Place it in the refrigerator for 30 minutes.

3. Cover the first layer of ladyfingers in the pan with the vanilla pudding from step 2.

4. Add another layer of coffee-dipped ladyfingers on top of the pudding.

5. Create the chocolate syrup: Maintaining the right ratio is crucial for achieving a crispy chocolate layer texture. The chocolate syrup ratio should be 4 tablespoons of sugar, 4 tablespoons of water, and 2 tablespoons of cocoa. After mixing these three ingredients, boil the mixture over medium heat. At the end, add the chocolate morsels. Once the morsels have melted, turn off the heat, add the butter, and wait until it melts.

6. Pour the chocolate syrup from step 5 evenly

over the second layer of ladyfingers in the pan.
7. Prepare whipped cream by using the heavy cream. Whip it twice and lightly cover the top of the dessert like a cloud. You can use a 2-cup glass measuring cup and an electric beater. No sugar is added to the whipped cream. It's recommended to use a cake pan with a plastic lid to keep the whipped cream safe.
8. You can skip step 7 by purchasing "Cool Whip", but be aware that it has a relatively high sugar content.

Traditional Korean Sweet Pastry (Yakgwa / 약과)

Total Time: 1 hour 30 minutes

Ingredient (4 persons):
- 1 cup of all-purpose flour
- 1 tablespoon of cinnamon powder
- 1 tablespoon of ginger juice
- 2 tablespoons of soju
  (a Korean alcoholic beverage)
- 1 teaspoon of Kosher salt
- 2 cups of canola oil (for frying)
- 3 tablespoons of roasted sesame oil
- 3 tablespoons of honey
- ¼ teaspoon of pepper
- 1 cup of organic sugar
- 1 cup of water
- 1 tablespoon of crushed pine nuts

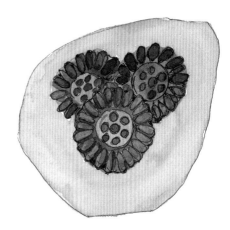

Method:

1. Mix all-purpose flour, cinnamon powder, and sesame oil thoroughly. Rub the mixture and then put it through a sieve.

2. Add ginger juice, soju, pepper, and salt to the mixture from step 1. Fold these ingredients together using a rubber spatula to form a firm dough.

3. Roll out the dough from step 2 to a thick height. Cut it into ½-inch tall, 1-inch by 1-inch square pieces.

4. Heat the canola oil to 300-320 degrees Fahrenheit and fry the pieces from step 3. Drain the excess frying oil and set them aside.

5. In a pot, combine sugar and water. Without stirring, boil it until you have 1 cup of syrup. Let it cool.

6. Dip the fried pieces from step 4 into the syrup from step 5 and sprinkle them with crushed pine nuts.

Well done! You've just completed making traditional Korean sweet pastries known as Yakgwa.

Twisted Cookies (Maejakgwa / 매작과)

Total Time: 1 hour 30 minutes

Ingredients (4 persons):
- 1 ½ cups of wheat flour
- 1 ½ tablespoons of grapeseed oil
- ⅓ cup of water
- 1 ½ tablespoons of organic sugar
- 1 ½ teaspoons of ginger juice (made from 1 ounce of ginger)
- ¾ teaspoon of Kosher salt
- 2 cups of canola oil (for frying)
- 1 cup of water (for making syrup)
- 1 cup of white sugar (for making syrup)
- 1 ½ tablespoons of pine nut powder
- 1 ½ tablespoons of cinnamon powder

Method:

1. Pour the grapeseed oil over the flour and mix thoroughly by rubbing it between your palms. Mix in the water, sugar, ginger juice, and salt. Knead the mixture well. Cover the dough with a wet cloth and let it stand for about 30 minutes.
2. In a pot, pour 1 cup of sugar and 1 cup of water, then slowly bring it to a boil without stirring. When the amount is reduced to half, let it cool. Add the cinnamon powder and mix well.
3. When the dough is smooth, roll it out into thin sheets and cut them into 1-inch by 2-inch rectangles. Slit each piece at the center without cutting to the ends. Pull one end of each piece through the slit to create a ribbon-like twist.
4. Deep-fry them over low heat (270 degrees Fahrenheit); otherwise, they may burn before they are fully cooked. After frying, soak them in the syrup from step 2. Before serving, sprinkle them with pine nut powder. Enjoy!

Glossary

measurement	volume	weight (metric system)
a pinch	1/8 teaspoon or less	
1 teaspoon	8~12 pinches	5 grams
1 tablespoon	3 teaspoons	15 grams
4 tablespoons	1/4 cup	60 grams
8 tablespoons	1/2 cup	120 grams
12 tablespoons	3/4 cup	180 grams
16 tablespoons 1 cup of water	1/2 pint	240 grams
2 cups of water	1 pint	
4 cups of water	1 quart	946 grams almost 1 kg
2 pints of water	1 quart	"
4 quarts of water	1 gallon	3,784 grams almost 4 kg
1 ounce	2 tablespoons	28.35 grams
16 ounces 1 pound	2 cups	450 grams

Reviewing Food & Nutrition

Carbohydrates – One of the 3 major energy sources in foods. The most common carbohydrates are sugars and starches. Carbohydrates yield about 4 calories per qram. Carbohydrates are found in foods in milk, vegetables, fruits, and starches.

- Starch – One of the 2 major types of carbohydrates. Starches are found in grains and beans.
- Sugars – One of the 2 major types of carbohydrates. Foods consisting mainly of naturally present sugars are those in milk, vegetable, and fruit. Added sugars include common table sugar and sugar alcohols (sorbitol, mannitol, etc.)

Protein – One of the 3 major nutrients in food. Protein provides about 4 calories per gram. Protein is found in foods from milk, meat, and meat substitutes. Smaller amounts of protein are found in foods from vegetables and starches.

Fat – One of the 3 major energy sources in food. A concentrated source of calories-about 9 calories per gram. Fat is found in foods in oil, meat, and meat substitutes. Some kinds of milk also have fat; some foods in starch also contain fat.
- Saturated fat – Type of fat that tends to raise blood cholesterol levels. It comes primarily from animal foods and is usually hard at room temperature. Examples of saturated fats are butter, lard, meat fat, solid shortening, palm oil, and coconut oil.
- Polyunsaturated fat – Type of fat that is usually liquid at room temperature and is found in vegetable oils. Safflower, sunflower, corn, and soybean oils contain the highest amounts of polyunsaturated fats. Polyunsaturated fats, such as corn oil, can help lower high blood cholesterol levels when they are part of a healthy diet.
- Omega-3 fat – Type of polyunsaturated fat found in fish and soybean oil known to lower triglyceride levels and protect the heart.
- Monounsaturated fat – Type of fat that is liquid at room temperature and found in vegetable oils, such as canola and olive oils. Monounsaturated fats can help lower high blood cholesterol levels when they are part of a lower-fat diet.
- *Trans* fatty acids – *Trans* fatty acids are fatty acids made through the process of hydrogenation, which solidifies liquid oils and the foods that contain them. These fatty acids tend to raise blood cholesterol like saturated fats do.

Mineral – Substance essential in small amounts to build and repair body tissue and/or control functions of the body. Calcium, iron, magnesium, phosphorus, potassium, sodium, and zinc are minerals.

Vitamins – Substances found in food, needed in small amounts to assist in body processes and functions. These include vitamin A, D, E, the B-complex, C, and K.

Nutrient – Substance in food necessary for life. Carbohydrates, protein, fats, minerals, vitamins, and water are nutrients.

Triglycerides – Fats normally present in the blood that are made from food. Gaining too much weight may increase blood triglyceride levels.

Fiber - An indigestible part of certain foods. Fiber is important in the diet as roughage, or bulk. Fiber is found in foods in starch, vegetable, and fruits.

Insulin – A hormone made by the body that helps the body use food. Also, a commercially prepared injectable substance is used by people who do not make enough of their own insulin.

Cholesterol – A fat-like substance found in blood. A high level of cholesterol in the blood has been shown to be a major risk factor for developing heart disease. Dietary cholesterol is found in all animal products. Eating foods high in dietary cholesterol and saturated fat tends to raise the level of blood cholesterol. Foods of plant origin such as fruits, vegetables, grains, and beans, peas, and lentils contain no cholesterol.

Alcohol – An ingredient in beverages, including beer, wine, liqueurs, cordials, and mixed or straight drinks. Pure alcohol yields about 7 calories per gram.

Index

Abalone Porridge (Jeonbokjuk / 전복죽) • 147

Acorn Jelly Making (Dotori Muk Ssugi /도토리묵쑤기) • 95

Baked Croaker (Jogi Gui / 조기구이) • 23

Baked Sweet Potatoes in Air Fryer • 259

Batter-fried Chives (Buchujeon / 부추전) • 24

Batter-fried Napa Cabbage (Baechujeon / 배추전) • 25

Batter-fried Stuffed Green Pepper (Putgochujeon / 풋고추전) • 26

Batter-fried Stuffed Mushrooms (Beoseot Jeon / 버섯전) • 27

BBQ Beef Short Ribs (LA Galbi / 엘에이갈비) • 28

Bean Curd Stew (Kongbijijjigae / 콩비지찌개) • 187

Bean Juice (Kongguk / 콩국) • 260

Beef Bone Soup (Sagol Guk / 사골국) • 189

Beef Braised in Soy Sauce (Jangjorim / 장조림) • 30

Beef Rice Bowl (Gogi Deopbap / 고기덮밥) • 148

Bibimbap (비빔밥) • 149

Black Bean Sauce Noodles (Jjajangmyeon / 짜장면) • 150

Black Beans with Soy Sauce (Kongjorim / 콩조림) • 96

Boiled Pickled Cucumber (Oi Sukjangajji / 오이숙장아찌) • 223

Braised Beef Short Ribs (Galbijjim / 갈비찜) • 31

Braised Bellflower Roots (Doraji Namul / 도라지나물) • 97

Braised Bracken (Gosari Namul / 고사리나물) • 98

Braised Burdock Root (Ueong Jorim / 우엉조림) • 99

Braised Chicken (Dakjjim / 닭찜) • 33

Braised Croaker (Jogijjim / 조기찜) • 34

Braised Fishcake (Eomuk Bokkeum / 어묵볶음) • 35

Braised Half-dried Pollock (Kodari Jjim / 코다리찜) • 36

Braised Lotus Roots (Yeongeun Jorim / 연근조림) • 100

Braised Mackerel (Godeungeo Jorim / 고등어조림) • 37
Braised Stalked Sea Squirts (Mideodeok Jjim / 미더덕찜) • 38
Brisket Vegetables Rice Soup (Gari Gukbap / 가리국밥) • 191
Broiled Pork Ribs (Dwaeji Galbi / 돼지갈비) • 40
Bulgogi (불고기) • 42
Cabbage Kimchi (Yangbaechu Kimchi / 양배추김치) • 224
Cabbage Rolls (Yangbaechu Roll / 양배추롤) • 43
Carrot Patties • 261
Chayote Pickle (Chayote Jangajji / 차요테장아찌) • 225
Chayote Salad (Chayote Muchim / 차요테무침) • 101
Chicken Corn Soup, Chinese Style • 193
Chicken Gravy • 102
Chinese Udong (Junghwa Junghwa Udong / 중화우동) • 152
Chives Kimchi (Buchu Kimchi / 부추김치) • 226
Chives Salad (Buchu Muchim / 부추무침) • 103
Cinnamon Ginger Punch (Sujeonggwa / 수정과) • 262
Cinnamon Spiced Sweet Potatoes • 264
Clear Stew with Codfish Head (Daegu Meoritang / 대구머리탕) • 194
Cold Cucumber Seaweed Soup (Oi Miyeok Naengguk / 오이미역냉국) • 104
Cold Gim Soup (Gim Guk / 김국) • 195
Corn Cream Soup • 196
Crabmeat Soup (Gesal Soup / 게살수프) • 197
Cranberry Sauce • 265
Cream Puffs • 266
Cucumber Pickle (Oiji / 오이지) • 227
Curry Rice (카레라이스) • 154
Deep-fried & Sugar Glazed Banana • 267
Deep-fried & Sugar Glazed Sweet Corn Ball • 269
Egg Roll Slices (Hwangbaekjidan / 황백지단) • 105
Eggplant Rice Bowl (Gaji Deopbap / 가지덮밥) • 155

Fermented Dry Squid with Rice (Bap Sikhae / 밥식해) • 106
Fermented Squid with Radish (Ojingeo Sikhae / 오징어식해) • 107
Fishcake Soup (Eomukttang / Odeng / 어묵탕, 오뎅) • 198
Fresh Baby Napa Kimchi (Eolgari Geotjeori / 얼가리겉절이) • 229
Fresh Baechu Kimchi (Baechu Geotjeori / 배추겉절이) • 231
Fresh Chives Kimchi (Buchu Geotjeori / 부추겉절이) • 233
Fresh Tomato Kimchi (Tomato Geotjeori / 토마토겉절이) • 234
Fried Cuttlefish (Ojingeotwigim /오징어튀김) • 44
Fried Gim with Rice Paper (Gim Bugak / 김부각) • 108
Fried Kelp (Dasima Twigak / 다시마튀각) • 109
Fruit Compote / Fruit Nut & Rice Salad • 271
Gimbap (김밥) • 156
Gim Mix (Gim Muchim / 김무침) • 110
Ginger Dressing • 111
Glutinous Rice Cake (Injeolmi / 인절미) • 272
Greek Yogurt with Cucumber & Apple • 112
Green Garlic Stem Dish (Putmaneuldaemuchim / 풋마늘대무침) • 113
Hamburger Patties • 46
Hand Rubbed Beef (Jumulleok / 주물럭) • 48
Hand Torn Dough Soup (Sujebi & Gamja Ongsimi / 수제비와 감자옹심이) • 199
Jellyfish Salad (Haepari Naengchae / 해파리냉채) • 115
Kimchi Stew (Kimchi Jjigae / 김치찌개) • 200
King Dumpling (Wang Mandu / 왕만두) • 49
Knife-cut Noodle Soup with Seafood (Kalguksu / 칼국수) • 158
Korean Ginseng Chicken Soup (Samgyetang / 삼계탕) • 160
Korean Patties (Wanjajeon / 완자전) • 51
Korean Style Dipping Sauce (Ssamjang / 쌈장) • 117
Korean Style Egg Roll (Gyeran Mari / 계란말이) • 52
Korean Sweet Dessert Pancakes (Hotteok / 호떡) • 274
Korean Sweet Potato Snack (Matang / 마탕) • 276

Korean Sweet Rice Cake with Dried Fruit and Nuts (Yak Sik / 약식) • 277

Kyung's Apple Salad • 118

Lasagna • 53

Mapo Tofu (Mapa Dubu / 마파두부) • 55

Meat Sauce for Spaghetti in Crock Pot • 57

Mixed Bone Soup (Seolleongtang / 설렁탕) • 201

Mixed Cold Buckwheat Noodles (Bibim Naengmyeon / 비빔냉면) • 161

Mochi • 278

Multi-grain Rice (Japgokbap / 잡곡밥) • 162

Mung Bean Pancake (Bindaetteok / 빈대떡) • 58

Namagashi (Saenggwaja / 생과자) • 279

New England Clam Chowder • 203

New Year's Day Rice Cake Soup (Tteokguk / 떡국) • 203

New York Style Clam Chowder • 164

Omurice (오므라이스) • 204

Organic Green Smoothie (Yachae Smoothie / 야채스무디) • 281

Oxtail Soup (Kkori Gomtang / 꼬리곰탕) • 205

Pan-fried Eggplant (Gaji Gui / 가지구이) • 119

Pan-fried Fish Fillet (Saegseonjeon / 생선전) • 60

Pan-fried Slice of Beef (Yukjeon / 육전) • 61

Parboiled Spinach Dish (Sigeumchi Namul / 시금치나물) • 120

Passover Broiled Beef Flank • 62

Pesto Sandwiches • 168

Pickled Green Pepper (Putgochu Jangajji / 풋고추장아찌) • 235

Pine-needle Rice Cake (Songpyeon / 송편) • 282

Pine Nut Porridge (Jatjuk / 잣죽) • 169

Pizzelle Cookies • 284

Platter of Nine Delicacies (Gujeolpan / 구절판) • 63

Pork Bulgogi (돼지불고기) • 65

Potato Pancake (Gamjajeon / 감자전) • 286

Pre-cut Kimchi (Mak Kimchi / 막김치) • 236
Radish Beef Soup (Mu Guk / 무국) • 206
Radish Kimchi (Kkakdugi / 깍두기) • 238
Radish Salad with Sugar & Vinegar (Mu Saengchae I / 무생채 I) • 121
Radish Salad without Sugar & Vinegar (Mu Saengchae II / 무생채 II) • 123
Raw Crab Marinated in Soy Sauce (Ganjang Gejang / 간장게장) • 124
Red Bean Sediment (Pat Anggeum / 팥앙금) • 287
Rice Cake Ball (Gyeongdan / 경단) • 288
Rice Punch (Sikhye / 식혜) • 289
Rice with Soybean Sprouts (Kongnamulbap / 콩나물밥) • 172
Rice with Kimchi & Soybean Sprouts (Kimchi Kongnamulbap / 김치콩나물밥) • 170
Rich Soybean Paste Stew (Cheonggukjang Jjigae / 청국장찌개) • 207
Salted Pollock Roe Stew (Myeongran Jjigae / 명란찌개) • 208
Sauerkraut • 242
Scallion Salad (Pamuchim / 파무침) • 126
Scones • 291
Seafood Dynamite • 66
Seafood Scallion Pancake (Pajeon / 파전) • 67
Seasoned Crab (Yangnyeom Gejang / 양념게장) • 127
Seasoned Cucumber & Bellflower Roots (Oi Doraji Muchim / 오이도라지무침) • 129
Seasoned Dried Pollock (Bukeopo Muchim / 북어포무침) • 68
Seasoned Dried Radish (Mu Mallengi Muchim / 무말랭이무침) • 130
Seasoned Mung Bean Sprouts (Sukju Namul Muchim / 숙주나물무침) • 131
Seasoned Shredded Squid (Ojingeo Chae Bokkeum / 오징어채볶음) • 69
Seasoned Soybean Paste (Gang Doenjang / 강된장) • 132
Seasoned Soybean Sprouts (Kongnamul Muchim / 콩나물무침) • 133
Seaweed Soup (Miyeok Guk / 미역국) • 210
Sesame Dried Pollock Soup (Bugeotguk / 북엇국) • 209
Soft Tofu Stew (Sundubu Jjigae / 순두부찌개) • 211
Sorghum Balls (Susu Pat Danji or Susu Gyeongdan / 수수팥단지, 수수경단) • 293

Soybean Paste Soup (Doenjangguk / 된장국) • 213
Soybean Paste Stew (Doenjang Jjigae / 된장찌개) • 214
Spicy Beef & Vegetable Soup (Yukgaejang / 육개장) • 215
Spicy Braised Chicken (Dakdoritang / 닭도리탕) • 70
Spicy Braised Tofu (Dubu Jorim / 두부조림) • 71
Spicy Korean Acorn Noodles (Bibim Dotori Guksu / 비빔도토리국수) • 174
Spicy Seafood Noodle Soup (Jjamppong / 짬뽕) • 176
Spicy Stuffed Cucumber Kimchi (Oisobagi / 오이소박이) • 240
Steamed Egg (Gyeran Jjim / 계란찜) • 72
Steamed Eggplant (Gaji Namul / 가지나물) • 134
Sticky Rice Cake, Covered by Red Beans (SiruTteok / 시루떡) • 295
Stir-fried Anchovy (Myeolchi Bokkeum / 멸치볶음) • 73
Stir-fried Chayote (Chayote Bokkeum / 차요테볶음) • 135
Stir-fried Dried Pollock (Bukeopo Bokkeum / 북어포볶음) • 74
Stir-fried Dry Squid Slice (Mareun Ojingeo Bokkeum / 마른오징어볶음) • 75
Stir-fried Eggplant (Gaji Bokkeum / 가지볶음) • 136
Stir-fried Eggplant and Anchovy (Gaji Myeolchi Bokkeum / 가지멸치볶음) • 137
Stir-fried Glass Noodles with Vegetables (Japchae / 잡채) • 76
Stir-fried Glass Noodles with Bell Peppers (Pimang Japchae / 피망잡채) • 78
Stir-fried Glass Noodles with Mushrooms (Beoseot Japchae / 버섯잡채) • 80
Stir-fried Golden-haired Squid (Ojingeo Silchae Bokkeum / 오징어실채볶음) • 82
Stir-fried Jiri Anchovies (Jiri Myeol Bokkeum / 지리멸볶음) • 83
Stir-fried Octopus (Nakji Bokkeum / 낙지볶음) • 84
Stir-fried Oyster Mushrooms (Neutaribeoseot Bokkeum / 느타리버섯볶음) • 86
Stir-fried Pork belly (Samgyeopsal Bokkeum / 삼겹살볶음) • 87
Stir-fried Radish (Mu Namul / 무나물) • 138
Stir-fried Red Pepper Paste (Bokkeum Gochujang / 볶음고추장) • 139
Stir-fried Rice Cake (Tteokbokki / 떡볶이) • 88
Stir-fried Soybean Sprouts (Kongnamul Bokkeum / 콩나물볶음) • 140

Stir-fried Zucchini (Hobak Bokkeum / 호박볶음) • 141
Stir-fried, Yellow-dried Pollock (Hwangtaegui / 황태구이) • 89
Sushi Rice • 178
Sweet and Sour Pineapple Pork (Tangsuyuk / 탕수육) • 90
Sweet and Sour Wings • 92
Tiramisu / Ladyfinger Coffee Cake • 297
Traditional Korean Sweet Pastry (Yakgwa / 약과) • 299
Tuna Sashimi Rice Bowl (Hoedeopbap / 회덮밥) • 179
Twisted Cookies (Maejakgwa / 매작과) • 300
Udong (Udong / 우동) • 181
Vegetable Soup (Yachae Soup / 야채수프) • 217
Vinegary Cucumber Side Dish (Oi Chomuchim / 오이초무침) • 142
Vinegary Lotus Roots (Yeongeun Chojeolim / 연근초절임) • 143
Watery Cabbage Kimchi (Yangbaechu Mul Kimchi / 양배추물김치) • 243
Watery Cold Buckwheat Noodles (Mul Naengmyeon / 물냉면) • 182
Watery cucumber Kimchi (Oi Mul Kimchi / 오이물김치) • 244
Watery Radish Kimchi (Mu Mul Kimchi / 무물김치) • 246
Watery Sedum Kimchi (Dolnamul Mul Kimchi / 돌나물물김치) • 247
White Kimchi (Baek Kimchi / 백김치) • 248
White Radish Pickle (Tongdak Jipmu / 통닭집무) • 144
Whole Kimchi (Baechu Kimchi / 배추김치) • 250
Whole Radish Kimchi (Chonggak Kimchi / 총각김치) • 252
Wild Pollock Stew (Maeuntang / 매운탕) • 218
Young Radish Kimchi (Yeolmu Kimchi / 열무김치) • 254

Searching in Korean by Korean Alphabetical Orders

가리국밥 191
가지구이 119
가지나물 134
가지덮밥 155
가지멸치볶음 137
가지볶음 136
간장게장 124
갈비찜 31
감자전 286
강된장 132
경단 288
게살수프 197
계란말이 52
계란찜 72
고기덮밥 148
고등어조림 37
고사리나물 98
구절판 63
김국 195
김무침 110
김밥 156
김부각 108
김치찌개 200
김치콩나물밥 170
깍두기 238
꼬리곰탕 205
낙지볶음 84

느타리버섯볶음 86
다시마튀각 109
닭도리탕 70
닭찜 33
대구머리탕 194
도라지나물 97
도토리묵수기 95
돌나물물김치 247
떡국 164
떡볶이 88
돼지갈비 40
돼지불고기 65
된장국 213
된장찌개 214
두부조림 71
마른오징어볶음 75
미탕 276
마파두부 55
막김치 236
매운탕 218
매작과 300
멸치볶음 73
명란찌개 208
무국 206
무나물 138
무말랭이무침 130
무물김치 246

물냉면 182
무생채 121 & 123
미더덕찜 38
미역국 210
밥식해 106
배추겉절이 231
배추김치 250
배추전 25
백김치 248
버섯잡채 80
버섯전 27
볶음고추장 139
부추겉절이 233
부추김치 226
부추무침 103
부추전 24
북엇국 209
북어포무침 74
북어포볶음 74
불고기 42
비빔냉면 161
비빔도토리국수 174
비빔밥 149
빈대떡 58
사골국 189
삼겹살볶음 87
삼계탕 160

생과자 279
생선전 60
쌈장 117
설렁탕 201
송편 282
수수팥단지, 수수경단 293
수정과 262
수제비 199
숙주나물무침 131
순두부찌개 211
시금치나물 120
시루떡 295
식혜 289
야채스무디 281
야채수프 217
약과 299
약식 277
양념게장 127
양배추김치 224
양배추롤 43
양배추물김치 243
어묵볶음 35
어묵탕, 오뎅 198
얼가리겉절이 229
연근조림 100
연근초절임 143
열무김치 254
엘에이갈비 28
오므라이스 154
오이도라지무침 129

오이물김치 244
오이미역냉국 104
오이소박이 240
오이숙장아찌 223
오이지 227
오이초무침 142
오징어식해 107
오징어실채볶음 82
오징어채볶음 69
오징어튀김 44
완자전 51
왕만두 49
우동 181
우엉조림 99
육전 61
육개장 215
인절미 272
잡곡밥 162
잡채 76
잣죽 169
장조림 30
전복죽 147
조기구이 23
조기찜 34
주물럭 48
중화우동 152
지리멸볶음 83
짜장면 150
짬뽕 176
차요테무침 101

차요테볶음 135
차요테장아찌 225
청국장찌개 207
총각김치 252
카레라이스 154
칼국수 158
코다리찜 36
콩국 260
콩나물무침 133
콩나물밥 172
콩나물볶음 140
콩비지찌개 187
콩조림 96
탕수육 90
토마토겉절이 234
통닭집무 144
파무침 126
파전 67
팥앙금 287
풋고추전 26
풋고추장아찌 235
풋마늘대무침 113
피망잡채 78
해파리냉채 115
호떡 274
호박볶음 141
황백지단 105
황태구이 89
회덮밥 179

Acknowledgement

"I extend my heartfelt thanks to the following contributors: Nansook, my dear friend, who has been a constant source of encouragement and advice; Stacy, for her invaluable artistic guidance from the beginning, including the idea of the beautiful cover design; my niece, Gayoung, for her hard work and sacrifice; David, for his valuable advice and guidance; and Q, my patient and understanding husband, without whom this book wouldn't be possible. I also want to express my gratitude to my kids—Sarah, Shin, X-tina, and David—for their unwavering support, and special thanks to my wonderful grandchildren—Jordan, Hansen, and Mason—whose joy completes this journey. Thank you all for making this book a reality!"

Artist: Kyung Shin
Title: Mountain, Deogyu
Size: 9"×24"
Medium: watercolor
North East Watercolor Society 39th Annual International Exhibition-2015, Kent, NY-exhibitor

Reference

1. Cho, E. J. (2008). Comfortable, Delicious, and Warm Dining Table like Mom's. Joongu, Seoul. Jayoon Printing.
2. Choi, B. S. (2007). Textbook of Soup Making for Daehanminguk Aunts. Jonrogu, Seoul. Woongjinsinkbig Inc.
3. Hepinstall, H. S. (2005). *Growing up in a Korean Kitchen*. Berkeley, CA. Ten Speed Press.
4. Hong, J. S. (2003). High-end Korean Food. Jongrogu, Seoul. Kyomunsa.
5. Bae, Y. H. (2003). Porridge of Korea with Wisdom of Ancestors. Jonrogu, Seoul. Hanrim Publishing Co.
6. Suzuki, T. (2003). *Japanese Homestyle Cooking*. New York, NY. Oxford University Press.
7. Hill, S. (2003). *Chicken over 400 Fabulous Recipes for All occasions*. Singapore. Barnes & Noble's Inc.
8. Nam, K. H. (2001). Classy Dining of Korean Food by 70 years Hand Taste of Grandmother, Nam Kyung Hee. Yongsangu, Seoul. Seoul Culture Inc.
9. Lee, H. B. (2000). Lee Hyang Bang Chinese Cooking. Seochogu, Seoul. Jubush.
10. Park, H. S. (2000). RECOMMENDED DIETARY ALLOWANCES for KOREANS. *7th Revision*. Korean Nutrition Society. Jung Ang Culture.
11. Jun. H. J. (2000). Traditional Korean Food. Sungdonggu, Seoul. Daewon Culture Inc.
12. Chang, S. Y. (1997). *A Korean Mother's Cooking Notes*. Sudaemungu, Seoul. Ewha Woman's University Press.
13. Wernert, S. J. (1997). *Live Longer Cookbook 500 Delicious Recipes for Healthy Living*. New York, NY. The Reader's Digest Association, Inc.
14. The Land O'Lakes Test Kitchens. (1992). *Treasury of Country Recipes*. East Montreal, Canada. Cy DeCosse Inc.